T0331892

Biomedical Ethics Reviews • 1990

Biomedical Ethics Reviews

Edited by

James M. Humber and Robert F. Almeder

Board of Editors

Biomedical Ethics Reviews • 1990

Edited by

JAMES M. HUMBER
and ROBERT F. ALMEDER

Georgia State University, Atlanta, Georgia

Humana Press • Clifton, New Jersey

The Library of Congress has cataloged this serial title as follows:

Biomedical ethics reviews—1983-   Clifton, NJ: Humana Press, c1982-

v.; 25 cm—(Contemporary issues in biomedicine, ethics, and society)
Annual.
Editors: James M. Humber and Robert F. Almeder.
ISSN 0742–1796 = Biomedical ethics reviews.

1. Medical ethics—Periodicals. I. Humber, James M. II. Almeder, Robert F.
III. Series.

[DNLM: 1. Ethics, Medical—periodicals. W1 B615 (P)]

R724.B493   174'.2'05—dc19   84-640015
                                              AACR 2   MARC-S

Contents

**Are the NIH Guidelines Adequate
for the Care and Protection of Laboratory Animals?**

Preface

Biomedical Ethics Reviews • 1990 is the eighth volume in a series of texts designed to review and update the literature on issues of central importance in bioethics today. Two topics are discussed in the present volume: (1) Should the United States Adopt a National Health Insurance Plan? and (2) Are the NIH Guidelines Adequate for the Care and Protection of Laboratory Animals? Each topic constitutes a separate section in our text; introductory essays briefly summarize the contents of each section.

Bioethics is, by its nature, interdisciplinary in character. Recognizing this fact, the authors represented in the present volume have made every effort to minimize the use of technical jargon. At the same time, we believe the purpose of providing a review of the recent literature, as well as of advancing bioethical discussion, is admirably served by the pieces collected herein. We look forward to the next volume in our series, and very much hope the reader will also.

James M. Humber
Robert F. Almeder

Contributors

Harold W. Baillie • Department of Philosophy,
University of Scranton, Scranton, Pennsylvania

James S. DeLaney • School of Public Health,
Harvard University, Cambridge, Massachusetts

Charles J. Dougherty • Center for Health Policy
and Ethics, Creighton University, Omaha, Nebraska

Arthur J. Dyck • The Divinity School,
Harvard University, Cambridge, Massachusetts

Leonard M. Fleck • Center for Ethics and Humanities
in the Life Sciences, Michigan State University,
East Lansing, Michigan

Thomas M. Garrett • Department of Philosophy,
University of Scranton, Scranton, Pennsylvania

William B. Irvine • Department of Philosophy,
Wright State University, Dayton, Ohio

Bernard E. Rollin • Department of Philosophy,
Colorado State University, Fort Collins, Colorado

Lilly-Marlene Russow • Department of Philospohy,
Purdue University, West Lafayette, Indiana

Should the United States
Adopt a National Health Insurance Plan?

Introduction

The first three articles in this section deal with the issue of justifying a national policy designed to provide universal health care for all citizens and residents of the United States. The first article is by Harold Baillie and Thomas Garrett, who begin their discussion by distinguishing between health care insurance and health care plans. Insurance is a mechanism by which economic risk for medical care is spread over a group; plans, on the other hand, involve the systematic delivery of health care and can be funded by specific or general tax revenues. Although Americans have traditionally favored insurance, Baillie and Garrett contend that a certain type of national health care plan is ethically necessary, and should be adopted in the US.

In defense of the view that we should adopt a particular health care plan, Baillie and Garrett argue that health care is a social good that acts by service to human dignity. Sharing in the social good identifies one as a member of society; thus, the social good must be available to all members of society. To serve human dignity, a system of health care should ease suffering and seek to return the ill to participation in society. When patients are hopelessly ill, they still deserve care, for they are members of society and share in the social good. However, pointless treatment is not called for.

Baillie and Garrett realize that health care competes with other commitments on which society must expend resources. Thus, they admit that it is not feasible for the govermnent to provide the highest quality health care for all citizens, and instead, claim that the principal concern for any health care plan should be to ensure adequate health care for all members of society. Baillie and Garrett contend that this goal is feasible, if defined within the context of

3

maintaining social participation. However, before any specific health care planning can occur, much "foundation laying" must take place. First, the professions must take the lead in providing a definition of "adequate health care" that distinguishes clearly between adequate and quality health care. Further, government bodies must stimulate public discussions so as to elicit majority support for a health care plan that guarantees adequate health care for all. Only after "adequate health care" has been defined and a public consensus has been reached concerning the desirability of ensuring adequate health care for all should planning begin for a system of national health care.

Baillie and Garrett assert that the social responsibility of providing adequate health care for all should not be taken to preclude members of society from seeking quality health care by entering the market. However, they insist that if the govermnent allows this option, it must take steps to guarantee that the market does not undermine the national health care plan.

The second article on national health insurance is by Charles Dougherty. Dougherty articulates the moral case for establishing a national health insurance program where national health insurance is understood as constituting a middle ground between socialized medicine and the free market distribution of health insurance.

Dougherty begins by noting the failures of our present health care delivery system. He then offers three moral arguments to justify instituting a national health insurance program. First, Dougherty argues that utilitarianism supports such a program, and that acceptance of utilitarian reasoning leads one to advocate national health insurance coverage that will allow the greatest number of people to lead long lives lived well. Second, Dougherty considers contractarian arguments and concludes that contractarians will support a national health insurance program that covers more than a utilitarian-based program. Rather than seeking to ensure a long life lived well for the greatest number of individuals, contractarians will try to structure a program that makes the least life among us the best it can be. Finally, Dougherty offers an egalitarian argument in support of national health insurance, and concludes that national health insurance can be justified as a middle ground between socialized medicine and the marketplace model of health insurance coverage.

After presenting the moral arguments for a national health insurance program, Dougherty considers libertarian and practical objections to a system of national health care insurance. Dougherty rejects all such criticisms. In the end, though, he allows that any acceptable national health insurance program must have limits on coverage. Dougherty admits that it will be difficult to set limits that restrict treatment. However, he argues that under the present health care delivery system, limits are being set in corporate board rooms without any inspection of the moral arguments for limitations, and without any consensus in a democratic community. As Dougherty sees things, a national health insurance program would improve on the present system, for it would bring debate into the public realm and help to achieve a consensus concerning the proper limits of insurance coverage.

The third article dealing with the issue of national health insurance is by William B. Irvine. Irvine believes that there is a great deal of support for some form of national health insurance in the United States today, and that this support arises because of the current crisis in American health care. Irvine does not see the crisis in health care as being one of quality, but rather one of cost. In the past several decades, medical costs have soared, and there is every reason to believe that they will continue to increase at an alarming rate. Working Americans and employers are being required to contribute greater and greater sums to employer sponsored health care plans, and approximately 37 million Americans have no health insurance at all. To ameliorate this crisis, many look to the federal government, arguing that the time is right for the institution of a national health insurance plan that will not only hold prices down, but also provide coverage for all citizens of the US.

Irvine opposes the view that national health insurance can serve as a solution to the current crisis in American health care.

Indeed, Irvine argues that the federal government is a causal factor in the present health care crisis, and that instituting a national health insurance plan could well *increase* the costs of medical treatment in the United States. Furthermore, if proponents of national health insurance were to argue for their position by claiming that national health insurance should be instituted even if it is not cost efficient because (1) the poor are entitled to health care, and (2) national health insurance would provide medical care for the poor, Irvine has two replies. First, Irvine says that he is not totally convinced that the poor are *entitled* to health care, and second, if the poor do have such an entitlement, Irvine argues that it would be more cost effective to provide them with vouchers that they could then use to purchase health coverage from private insurance companies, rather than to allow the federal government to direct a health insurance program

After arguing that national health insurance will not resolve the problems inherent in the current national health care crisis, Irvine examines what he takes to be the causes of that crisis. This examination leads Irvine to argue that the best way to contain the ever increasing costs of health care is to destroy medical monopolies and increase competition among health care workers. To achieve this goal, Irvine would "blur" the lines that currently demarcate the various professions, and allow "technicians" with limited but specialized spheres of knowledge to do things that they presently are not permitted to do (e.g., Irvine would allow some technicians to perform simple surgery, even though those technicians did not possess an MD degree). In Irvine's view, by allowing technicians to compete with physicians, and by allowing currently recognized health care professionals (e. g, nurses, pharmacists, and physicians) to compete with one another in certain areas of medical stewardship, we would operate to keep the costs of medical care to a minimum.

The final two articles on the topic of national health insurance proceed by critically evaluating specific proposals for national health insurance coverage. The first of these articles is by

Arthur Dyck and James DeLaney. By examining specific health care proposals, Dyck and DeLaney hope to offer an ethical assessment of the principal arguments for and against a national policy designed to provide universal health care insurance within the United States. In the process, they also attempt to determine why the United States stands alone among major industrial countries without comprehensive and universal health care coverage for all of its citizens and residents.

Dyck and DeLaney proceed in their assessment by examining four specific proposals for national health insurance. These are

1. the Himmelstein-Woolhandler proposal;
2. the Uwe E. Reinhardt proposal;
3. the Enthoven-Kronick plan; and
4. the health insurance system proposed by Rashi Fein.

When evaluating these proposals for national health insurance, Dyck and DeLaney do so from four perspectives. First, they ask how each proposal attempts to satisfy the demands of justice. Second, they try to determine whether each of the proposals is in any sense necessary. Third, they evaluate the probability of success for each plan. And finally, they ask how cognitively sound each proposal is.

After examining the above four proposals for national health insurance, Dyck and DeLaney draw a number of conclusions and offer their own suggestions for formulating a national health insurance policy. First, they note that all four of the above proposals agree that

1. the United States should not follow England in establishing a system of socialized medicine;
2. health care should not be left to the vagaries of an unbridled free market; and
3. the federal government should play some role in any system of national health care.

Dyck and DeLaney appear to accept these views. Further, they conclude that the principal reason the United States does not have a national health insurance plan is not that there is deep-seated disagreement about the need for such a plan. Rather, debate about national health insurance has centered almost exclusively on specific proposals, and has lacked wide public involvement; thus, when a specific plan is advanced, that proposal lacks mass appeal and is opposed by proponents of alternate national health insurance plans. To avoid such occurrences in the future, Dyck and DeLaney accept two suggestions made by Rashi Fein. Specifically, they agree with Fein that (1) debate should not center on concrete proposals for national health insurance, but rather on the broad principles that should guide national health insurance policy, and (2) we should ensure greater public participation in the debate concerning national health insurance, and increase public education in this area.

All four of the proposals for national health insurance that Dyck and DeLaney evaluate agree that the federal government should enforce cost containment. Dyck and DeLaney reject this view. They argue that we are now faced with a crisis in health care—people are dying in preventable ways—and that a national health insurance plan is needed to combat this crisis. Moreover, to make the case for a national health insurance, one must show that adoption of a plan will decrease rather than increase discrimination, and increase rather than decrease the quality of health care. Neither goal can be achieved if national health insurance plans require that the federal government practice cost containment. What is required, Dyck and DeLaney argue, is not cost containment, but rather responsible, honest, nonwasteful expenditures for health care, together with public scrutiny to bring inflationary pressures under control.

In the end, Dyck and DeLaney hold that we should recognize a right to health care. Recognition of this right is required for the protection of human life. The protection of human life is not something individuals can accomplish by themselves; it must be a communal effort. Hence, Dyck and DeLaney argue, "the kind of communal protection of lives provided through armed services, police forces, and fire departments [should] be extended to health care...[for] what can be claimed as rights and affirmed as communal responsibilities in any of these activities can be similarly claimed and affirmed in all of them" (pp. 122–123).

Although they do not say so explicitly, Dyck and DeLaney appear to believe that the only proposals for national health insurance that are feasible in the United States today are those that embody features inherent in the moral, political, and economic values found in our country's history and political culture. Leonard Fleck disagrees with this view, and argues instead for a national health insurance plan that he claims breaks with traditional economic and political values.

Fleck begins by listing six features that he believes are commonly found in "feasible" national health insurance proposals. First, proposals of this sort reject a single mechanism for financing health care. Second, they advocate a multitiered health care system. Third, they propose a mixed, public/private system for financing health care. Fourth, they guarantee only minimally decent care for all citizens. Fifth, they share a belief that competition is the best means for keeping medical costs to a minimum. And finally, "feasible" proposals for national health insurance reject centralized decision-making on resource allocation by government bureaucrats, and instead hold that allocation decisions should be made by individual consumers or organized groups of providers at the local level.

Fleck argues that all proposals for national health insurance that exhibit the above features are flawed, and that the best available health care plan is one that is similar to the comprehensive, government operated system now available in Canada. To support this view, Fleck offers two competing health care proposals for evaluation. The first proposal is the Enthoven-Kronick plan for national health care insurance. This plan is broadly libertarian in its conception of justice, and is judged by Fleck to be the best proposal embodying the six features listed above as characteristics of "feasible" health care plans. The second proposal is the national health insurance plan offered by Himmelstein and Woolhandler. This plan mirrors the Canadian system of financing health care, and rejects the six features exhibited in the Enthoven-Kronick proposal. After outlining both proposals, Fleck evaluates each, using what he calls a "nonideal framework of justice." In the end, Fleck concludes that the latter proposal is rationally preferable to the former from a variety of value perspectives.

The Ethics
of Social Commitment

Harold W. Baillie
and Thomas M. Garrett

The Problem

The Context

A national health care solution is inevitable for the US. There are thirty-five million uninsured Americans, a third of whom are employed. The increase in medical costs is hurting the middle class as medical plan costs are shifted onto their paychecks by their employers. A large group who do not qualify for Medicaid are unable to pay for private insurance. The moral problem of those uninsured has been coupled with the economic problem of the middle class, and so, the business and political worlds have started to take heed.

In attempting to frame a national health care solution, it is helpful to distinguish between a health care plan and health care insurance. Insurance is a mechanism by which economic risk is spread over a group; in short, there is trade-off in which the individual shares the burdens of the group in return for the assistance the group offers the individual in time of need. Insurance demands specific contributions, generally on an actuarial basis, and provides specific benefits. Health care insurance does not necessarily involve systematic delivery of the health care.

From: *Biomedical Ethics Reviews • 1990*
Eds.: J. Humber & R. Almeder ©1991 The Humana Press Inc., Clifton, NJ

Plans, on the other hand, involve the systematic study and delivery of health care, but need not involve specific economic contributions. Since benefits are usually specified, plans offer limited benefits equally to all members. Plans need not be funded by actuarially based contributions, but can be funded by either specific or general tax revenues. The advantages involve the equality and universality of the offered care; the disadvantages are the loss of choice and the absence of market influence on quality.

Traditionally, Americans have favored insurance, in part because it is an economic means to a solution that preserves the primacy of choice and uses market forces to control quality. It has turned out, however, that the current method of health care delivery in the US has not seen the full expected advantageous effects of market forces, particularly on costs. Insulated by third-party payers, by ethical concerns with the confidentiality of medical records, and even by the esoteric knowledge characteristic of health care itself, patients in the guise of consumers have been prevented from exercising judgment based on a reasonable evaluation of fully disclosed information. Ignorant consumers cannot use the options offered by market to control either their medical costs or the quality of the care they receive.

One further, and important difference between insurance and planning is that insurance is an economic strategy and planning is not. That is, one buys insurance because it makes economic sense, not because it is the (morally) right thing to do. Thus, both insurers and insurees legitimately include economic considerations in the formulation, maintenance, and purchase of insurance. There are incentives to exclude certain groups of people that might increase the economic risks of the insured group. For example, a health insurance group may exclude anyone over forty-five or anyone who smokes because these characteristics increase the economic risks for the whole group. Although it can be mandated that an insurance group not behave this way, or even that it should take in only smokers or the elderly, such redirection would come from outside the insurance group.

In other words, a national health plan and national health insurance are not incompatible issues. Any insurance mechanism will require, or at least be compatible with planning, whereas planning will require a source of funding, be it insurance or taxes. Rather than begin with a debate regarding the superiority of insurance or planning, it is more instructive to begin with a discussion of what national health care should do.

This discussion of funding health care indicates part of the context within which the problem of national health care is situated. The ethical problems of national health care are, largely, only subproblems of the ethics of distribution and the rationing of health care. Thus, we will argue that when the full nature (both its foundation and its limits) of a fair distribution of medical care is understood, it will be clear that a certain type of national health care plan is ethically necessary. Our considerations will raise questions about the nature of fairness as well as the legitimate ends of health care. Finally, since ethics is ultimately aimed at practice, the feasibility of a plan is a prime concern.

Clearing Some Brush

There are a few concerns that must be kept in mind in order to properly begin this discussion. First of all, we are talking about an area in which there is a shortage of resources. Distribution and rationing of health care are necessary because there are more possible medical needs than there are resources to meet those needs. Further, there are other areas that assert competing claims for these scarce resources. No one is suggesting that any of these areas, including health care, be the sole concern of government or of society. Thus, there would be apportionment of resources, as well as rationing.

Another concern is the idea of quality of care. As we will argue, the rhetoric of quality care in medicine is damaging to reasonable evaluations of what a health care system should do. It blurs the distinction between the cutting edge of medical care, which can be available only to a few, and the routine abilities of medical care, which can be much more widely available. The cutting edge is not only expensive, but it is sustained by the very talented and driven few who push beyond what is known. This kind of treatment cannot be widely available by its very nature. As techniques become settled, as production catches up with technological advances, and as new skills become the substance of health care education, what was the cutting edge becomes routine and, now by its nature, more widely available.

Routine health care can be done well, or it can be done shoddily. It is part of the ethics of the health care professions that health care should be delivered well. In this sense of quality health care, everyone does deserve quality care, and ceaseless efforts, particularly by the professions, should be made to provide it. But when quality health care refers to the cutting edge of medical care, and when it is assumed that all should have access to cutting edge health care, then we are talking about a practical and ethical impossibility.

Finally, we need to remember that we are discussing a political issue that, by its nature, encompasses many interest groups and concerns. Some groups are at this time exercising great control over health care delivery, whereas other groups, which perhaps should, are not. Also, a group's interests may be divided, sometimes in ways that can seriously impair its ability to function. For example, the American Medical Association remains both politically powerful and worried about any possible loss of control, both of income and medical practice.

Ethics of Distribution: How Is Fair Distribution Determined?

Dignity and Society

Ethicians have a responsibility to develop a sense of fair distribution that is responsive to the variety and relative importance of necessary social goods. The definition of fair distribution is not primarily the responsibility of the medical or political professions. Certainly, both professions must contribute to the discussion, but they bring expertise that does not directly address the question of distribution. It is ethics that addresses such questions of fairness independently of biases of medical specialty or political bases, and it is ethics that can make use of a tradition of reflection on fair distribution. At the same time, political discussion is crucial in specifying the content of many duties.

Consonant with a long and powerful tradition in Western civilization, contemporary medical ethics focuses on the dignity or worth of the individual. Although there are highly abstract discussions of the meaning of dignity, it is important to note that when it comes to using the concept in practical matters, dignity is unavoidably understood in a social context, determined by the history and material conditions of the society, and by the social roles, imaginative hopes, and experience of the individual.

The social nature of dignity embodies the interdependence of the individual and society. Since, for the sake of this discussion, individuals exist only in the context of society, and thus, have dignity only in a social context, issues fundamental to maintaining an individual's dignity are basic and universal concerns for the individual when confronting society (frequently put in terms of rights with claims against society). On the other hand, the society depends on the individual for its existence. One of the basic political insights of modernity (and the foundation for rights talk) has been the recognition that the power of the state can submerge the individual, leaving the state to exist simply for the sake of power. This is as destructive of the state as it is of the individual, so the society must take as its basic concern the dignity of its members (e.g., protect the rights of its citizens).

Such basic and universal concerns ground actions that are socially necessary for the satisfactory maintenance of the social fabric. No actions, however necessary they appear, can be justified when performing them would destroy the society, and through it, the individuals who compose the society. Thus, basic and universal concerns ground necessary, but not absolute ethical requirements for action.

The task of balancing the demands of dignity and the needs of society so as to maintain both the individual and society is accomplished by political wisdom. The exercise of political wisdom is a most difficult problem, and we will return to this issue at the end of the paper. For now, it must be emphasized that the tension between interests of the individual and the society is the heart of any attempt to talk of political wisdom. Thus, political wisdom can only be exercised by the entire group, even when it needs the leadership of some subgroup.

The economic market has often been used as an alternative to political wisdom, side-stepping the need to have leadership generate a consensus in the group.[1] Central to our evaluation of fair health care distribution is the claim that such market-based distributions are unfair and inefficient in the case of medical care. Medical care is not a typical economic good. Patients are drawn to medical care in crisis circumstances, when shopping around for the best deal is not possible. Further, patients are largely ignorant of the content and justification of medical care. Except in rare circumstances, we are not in a position to evaluate the scientific or practical basis of a medical treatment as we are to evaluate a car or washing machine, and the stakes are much higher. Finally, patients do not have access to full information about the records of physicians, hospitals, and so on, to which they go for medical care. Thus, evaluating quality on the basis of performance is not possible. Confidentiality and professional courtesy will continue to impede any efforts to break down these barriers to the free flow of information.

If the market cannot be used as a mechanism for establishing a fair distribution, we must in some way plan a fair distribution, reaching beyond the individual to some conception of how such a distribution ought to occur. Before attempting to sketch the rudiments of the principles of ethical distribution, we must first discuss the purpose of health care, since this, like the availability of resources, provides the realistic framework that forces a recognition of the limits on the human desire to help everyone.

What Is the Purpose of Health Care in General?

Health care is a social good, consisting of service to the dignity of the human person as it exists in a body that can die or suffer problems in form or function.

By a social good, we mean a good that cannot exist without the cooperative efforts of the society, and is the possession of the entire society. A social good is valuable in itself precisely as the result of this cooperative effort. It is also good for the opportunities that it makes available to members (generally understood as public goods). But as the result of a cooperative effort, a social good is among the highest accomplishments of social life; sharing in this good, in effect, identifies one as a member of this society. Thus, the social good must be available to all members of the society, or they are effectively forced out of the society.[2]

By service to human dignity, we mean that health care acts on behalf of human dignity, assisting the human person as it suffers the biological and psychological dysfunctions or deformations that are or come to be present in the human body. The specific nature of this service is developed as the arts of health care develop. The more we know and the greater our skills and technology, the more potential we have to serve human beings through care of their bodies and minds.

Participation in society refers to the ability of the individual to actively share in some form of social life. This need not be economic life, nor need it be life at a level defined by some requirement of human nature. As we suggest later, a precise understanding of "participation" will require the development of a social consensus, the struggle toward which is already well under way in hospitals and courts.[3]

Health care is an issue of justice because the individual may assert two types of claims against society: first, the right of sharing in a social good as confirmation of one's social identity; and second, a claim to society's assistance in remaining a member of that society (right of opportunity).[4] This service to the dignity of an individual provided by health care may be accomplished in two ways, either in the easing of suffering from irremediable causes, or where possible, in the remediation of dysfunction or disfigurement, where remediation means, at the very least, a return to participation in society. For example, a patient with a potentially fatal illness can expect medical care to avert that death as part of his share of the social good.

The distribution of scarce health care resources is unjust when it diminishes human dignity by preventing an individual or group of individuals from sharing in this social good, or by denying an individual aid when access to health care would reasonably be able to insure or dramatically improve their participation in society. At a minimum, this would include providing care (but not pointless treatment) for a hopelessly ill patient. Even though the patient's physical condition will not improve, as a member of society that individual can expect a reasonable share of this social good.

What Portion of Health Care
Should Be Supported by a National Health Care Plan?

The previous section provides the roots of a society's obligation to provide health care to its members. These roots by themselves do not and cannot provide the content of the obligation, that is, they do not provide a description of what health care is to be provided. The determination of the content is a political decision made in response to the ethical demands contained in these roots. In other words, the whole of society must decide and define this content in terms of its values and resources. A little later, we will speak of the various groups that must be involved in arriving at a consensus about this content; here, we wish only to sketch the political philosophy that is a necessary part of the ethics of distribution.

It is worth remembering that the limits of the society's commitment to provide this social good is the balance between the reasonable utility of the health care and the other commitments on which society must expend resources. The actual distribution of health care resources is a function of political wisdom, insofar as such distribution defines society's ability and willingness to maintain or restore an individual's participation in society, as well as what it means to participate within the larger context of the society's full range of responsibilities.

The individual's ethical commitment to the limits to health care generated by political wisdom arises out of the nature and use of social goods. The individual has a stake in these goods that grows out of the cooperative effort that produced them, and through them, produced part of the identity of the individual. Any actions that threaten the existence of these goods, such as misusing resources, threatens a good that is the individual's concern. A nonmedical example is patriotism. Since the individual needs the good of society and has contributed to the existence of this particular society, the existence of that society is good for the individual even to the extent that the individual might choose to sacrifice themselves for it. In health care, the individual can recognize the value of the health care system even as he recognizes that it can (reasonably) do nothing for him at this time.

In sum, within the feasibility limits to be discussed in the next section, society is obligated to ensure health care that accomplishes the task of maintaining an individual's participation in society, or when participation is not possible, caring for that dying person. This is the root of a distinction according to which rationing a social good can take place; the difference between adequate health care (necessary to maintain dignity) and quality health care (the provision of refinements in treatment that go beyond the basic activity of social participation).

Feasibility: What Can Be Done?

Limitations

To adequately evaluate the structure of a national health plan, we must keep such an effort in perspective. The feasibility of such a plan exists in the context of other types of contributions to the national health, e.g., housing and education. It is important to remember that acute health care, which has been the focus of most of the contemporary discussion within medical ethics, is only a part of the influences on health in a society. Public health measures, such as sewer systems and clean running water, have done more to improve the overall health of a society than most acute care measures. Also, discoveries such as antibiotics have a profound impact on well-being and can be made widely available with relatively little expense.

Pertinent also is the variety of claims that are legitimately made against society's resources. The feasibility of a health plan thus exists in the context of other demands on society's resources, e.g., defense, the economic infrastructure, and so on. A society is a complex institution with many responsibilities, including that of its own survival. As important as health care is to a sick individual, the impact of that individual's needs are mitigated by the competing needs of society. Since the dignity of the individual exists within the social environment, that environment cannot be destroyed for the sake of the dignity.

Finally, there are physical limitations that constrain any health care plan. Geographical location, the natural processes of aging and death, unavoidable tragic choices between goods, and chance occurances all conspire against human control and the establishment of full equity. For example, it will never be the case that we can provide the same quality and scope of medical care to the inhabitants of a small Montana town as we provide to those who live near major medical center.

Feasibility

Adequate health care can meet these limitations if defined within the context of maintaining social participation. Such standards can be developed as fruits of the political process with the health care professions leading the discussion (what was earlier called practical wisdom).

Health care that emphasizes delivery of the highest possible quality cannot function within these limitations. Pursuit of the highest quality will favor those who are already successfully participating in society and exclude those who are not (very often, this involves wealth, but educational and social status often substitute). This defeats the claim that health care is a social good, and it encourages a disregard for a prudent balancing of competing social goods.

Planning: An Exercise in Developing Political Wisdom

National Health Planning

Planning must begin with a recognition that the primary concern of a national health care plan is adequate health care for all citizens. The provision of adequate health care is a social obligation, not an economic entitlement. Thus, this is not an insurance proposal (the exercise of choice in conjunction with actuarial funding or the spread of risk), but socially funded commitment (based on ethical obligations intrinsic to society). Certainly, treatment beyond adequate care (to what we have called quality care) may be provided through a variety of schemes, such as insurance, but that type of treatment represents a luxury, the provision of which is not a social obligation.

This distinction between adequate and quality health care is the central issue in health care planning. It is also the central issue in understanding the ethics of health care as applied to several other important areas, such as termination of treatment in the process of dying. As our medical knowledge and skill increases, the gap widens between what we can do and what we should do in certain circumstances. Thus, it has become imperative to always remember that the goal of health care is service to the dignity of the patient; harsh experience is teaching us that medical treatment is not a goal in itself.

Defining a National Health Care Plan

Given the nature of the scientific, political, and experiential knowledge necessary to distinguish between adequate and quality health care, the professions in conjunction with society must be called on to define medical practice in a manner that provides for definition and provision of adequate care. Given the current practice of providing unlimited care, this task is unusual and has not as yet been done. The plan called for here requires a massive rethinking and restructuring of medical care, both how we think of it and how we deliver it.

The principal members of the health care field (physicians, nurses, physical therapists) are generally called professionals. As professions, these groups can be called on by society to take a more active role in defining their own practice and in relating that practice to social concerns and limits. Professions have always had special social responsibilities, the correlate of their privileges, and the acknowledgment of this task represents a more formal construction and exercise of professional responsibilities.

Responsibilities of Professionals

In concert, the health care professions must face the need to do more to regularize medical practice (be more conscious of the need to develop and enforce scientifically justifiable standards of treatment) and to be aware of the actual contribution to the best interest of patients and society made by new technologies and modes of treatment. Any implementation of the concept of adequate health care will require a reasonable standardization of treatments, and an accurate and continuing assessment of the usefulness of treatments. A system involving grades of treatment can only be fair when there is a constant effort to insure, insofar as possible, that we are all talking about the same options.

Health care professionals must also be willing to assert professional leadership in developing a clear public understanding of the capabilities and limits of the medical profession, and participate in discussions of the mutually dependent roles of all health care professions. Public concerns will control the success or failure of any health care system, and an informed public with realistic expectations can make possible the fair delivery of adequate health care. Public misconceptions can lead to unrealistic goals, false fears or hopes that can demand overtreatment, all of which contribute to an unmanagable, unethical health care system.[5]

The Mass Media

The discussions among professionals will not have full impact unless they are sold to the public. In practice, the opinions and desires of the public are as powerful as professional opinions in determining the shape of a health care system. To sell the new concepts, the informed cooperation of the mass media is necessary, if not sufficient. We say "informed cooperation" since, at present, the media are often content to report unvalidated treatments or experiments that never yield fruits. In some cases, they do no more than report the opinion of some physician or print summaries from a prestigious journal. In the process, they rarely follow up on stories, or attempt to put information in context. This results in a skewed public perception of possibilities and accomplishments of the health care system and an inflated sense of "quality care." What is needed is more intelligent reporting in which there is a sensitivity to context, with an informed appreciation of defects, omissions, or scientific limitations of the material.

In addition, the mass media should continue to discuss the costs of treatment and to continue to dramatize the crisis in health care financing. In so doing, it can ensure realism in its reporting of the possibilities of health care. The idea that given enough time and money we will have a cure for everything can easily invade popular media, yet it only obscures the real issues of the discussion.

Public Debate and the Responsibilities of Society

The dramatic shift in the conception and delivery of health care as outlined here requires this full participation. Society must generate the specification of social participation as the goal of adequate health care, and it must contribute to the definition of what is adequate to bring about this participation. This is because society remains the final judge of the validity and success of any such attempt, and without the cooperation and commitment of the majority of the group the endeavor will fail.[6] Examples as diverse as Prohibition and the public response to the Arab oil embargo suggest both the need for consensus and the possibilities for accomplishment when the consensus is achieved.

Mass media presentations are necessary but not sufficient for the education of the public. Public discussion is also necessary. By public discussion, we do not mean mass meetings or even lectures, but the full variety of means, both formal and informal, by which a social consensus is formed. From debates in bars and living rooms, struggles in professional organizations and labor negotiations, and the suffering of those sick and those caring for the sick, will emerge a sense of the real grounds on which consensus can be made.

Responsibilities of the Government

In the first phase, the government should do no planning that would preempt the formation of a public consensus. Such planning would likely build a model on outmoded cultural assumptions or on the concerns of special interest groups. The first duty of the government, and in particular the legislatures, will be to stimulate and keep running the public discussions we have described. Legislators should make discussion of alternate health care delivery systems a priority item. They should have these discussions with constituents of all classes, from the homeless to upper income groups.

Only when such a consensus has been reached will the government be justified in planning a national health care system.[7] Then, within the limits/guidelines established by the discussion, the government has an obligation to provide reasonable access to adequate health care for all its members.

Alternate Care

Quality health care beyond the adequate level may be obtained by entering the market. It is ethically possible, and economically and socially appropriate to allow individuals to purchase some form of additional medical coverage. The primary social responsibility of providing adequate care to all members of society does not preclude citizens from pursuing, on their own initiative, health care beyond what is to be provided by a national plan. This option for further care would respect the traditional reward system characteristic of our market economy, and allow individual choice to operate beyond the limitations of the national plan.

In order to prevent health care on the private market from undermining the national plan, all those involved would have to work out the relationship between the two options. A balance between professional support for the market and for the guaranteed adequate health care must be struck through government support for all aspects of the provision of adequate care, through a genuine public consensus as the foundation for adequate care, and through a joint effort with the professions in demystifying health care.

Although such effort would entail some difficulties, there are a couple of ways of doing this. One would be to have the government as an insurer for certain levels of care, so that physicians would bill the government for adequate care; private insurance could be purchased to cover other levels of care. Another possibility would be to have two systems, much like the primary and secondary educational system in the US, which has both private and public elements.

Conclusion

We have argued that it is ethically obligatory to develop a national health care plan that provides adequate health care to all citizens, regardless of ability to pay. The basic reason for this obligation is the correlation of human dignity and participation in society. The definition of what constitutes adequate health care must be worked out by society in a political discussion lead by health care professionals, but involving the widest variety of participants. Because of reality limitations, it is unlikely that a national health care plan, premised solely on the delivery of highest quality health care, could accomplish this ethical obligation.

It should be clear that our first steps toward the consideration of a national health care plan are fraught with difficulty. Some will say that our suggestions are impossible since they call for a rethinking of the entire problem, and present a challange to the rhetoric that has enshrined the idea of quality health care for all. If we cannot arrive at a social consensus that accepts reality limitations, we will end up with a plan that once again tries to do too much, and in the process, leaves large segments of the population without any health care and other groups with less than adequate care. In short, led by the Pied Piper of quality health care, we will be right back where we started, with an unethical system and with a larger health care bill than ever.

Notes and References

[1] *See,* as recent examples, Alain Enthoven and Richard Kronick, (1989) "A Consumer-Choice Health Plan for the 1990's, Part I" *New England Journal of Medicine* **320,** 29–37; and David U. Himmelstein

Steffie Woolhandler, and the Writing Committee of the Working Group on Program Design (1989) "A National Health Program For The United States," *New England Journal of Medicine* **320,** 94–101. The primary concern of these plans is fiscal, not ethical. They all presume competition will control both quality and costs of health care. An example of how the market does not function in health care can be found in the growth of hospitals in Fort Wayne, Indiana, as reported in "Medical Waste: Hospital Construction Booms, Driving Cost of Health Care Up," *The Wall Street Journal,* January 10, 1990, A1, A4.

[2] Rawls understands primary social goods to be those goods that society must provide to enable the individual to pursue their individual life-plans. That is, primary social goods are means (*Theory of Justice* Harvard University Press, Cambridge, MA, 1971, p. 93). We are attempting to show that social goods are not simply means to individual goods, but are also goods that have intrinsic social value themselves, conferring and acknowledging membership in a society, and thus, defining and specifiying the dignity of the individual. Rawls himself complicates matters with his later discussion of respect, but many interpreters, such as Norman Daniels in *Just Health Care* (Cambridge University Press, Cambridge, UK, 1985), apply to health care distribution only the issue of opportunity.

[3] For example, in the Cruzan case.

[4] *See* Note 2

[5] Given the current structure of the health care professions, physicians have a responsibility to lead the discussion to identify health care priorities consistent with a distinction between adequate and quality health care. There is, of course, room for concern in giving physicians this leading role (*see* Paul Starr, *The Social Transformation of American Medicine,* Basic Books, New York, NY, 1982; and Barbara Ehenreich and Dierdre English, *For Her Own Good: 150 Years of Experts' Advice to Women,* Anchor Books, Garden City, NY, 1979), but there seems to be little choice in this preference and some benefit. Physicians are at the center of the full diversity of health care options in the US. Their expertise is necessary for the development of a plan that scientifically and experientially will reflect the greatest extent of care and treatment possibilities.

In particular, this role for physicians flies in the face of recent concern with the past and present entrepeneurial role of the physician, and the recent collapse of the independent practices, emergency rooms, or health maintenance organizations. Indeed, the profession might not be up to it. But much of the concern expressed by physicians about the future of medical practioners centers around the loss of control in the face of changing forms of the delivery of medical care. If physicians are to remain professionals, then they must inform and contribute to that change. As the experts in the field, they are legitimately a significant part of the group that society may expect the most from in any development of the field.

Other health care professions, especially nursing and hospital administration, must also enter into the discussion of adequate health care and the cooperation of the health professions in delivering it, so that all professional interests, as well as all perspectives on patient interests, will be represented.

Nursing, in particular, is in a position to offer alternatives to regular medicine that could profoundly influence the outlines of adequate and quality health care. The recent development of the profession of nursing, including both educational and functional improvements, entail a more creative and independent stance by the profession.

⁶This is a partial response to a criticism made by Allan Buchanan that positions that argue for a decent minimum are "virtually contentless" (*see* "Health-Care Delivery and Resource Allocation," in *Medical Ethics,* edited by Robert Veatch, Jones and Bartlett, Boston, MA, 1989). Our position is that adequate care can only be defined by the society in light of its understanding of what can reasonably be done at that time to restore, if possible, a person to participate in society. The algorithm that correlates such plastic terms as "participation in society," "medical knowledge, skills, and technology," and "available resources," must be generated in the historical and material context. We think that our position has a clear purpose and method, but its content must be generated by a social consensus.

⁷The recent repeal of the *Catastrophic Health Care Act* shows how sensitive Congress needs to be with regard to establishing contact with all groups, not just self-appointed or official representatives.

The Moral Case for
National Health Insurance

Charles J. Dougherty

Introduction

In this article, the moral case for establishing a program of
national health insurance in the US will be articulated and defended. The phrase "national health insurance" is not meant to
refer to any specific program or proposal. It is meant generically
to refer to any arrangement for the financial pooling of health care
risks that provides universal basic health insurance coverage to
all Americans and is operated at least in part by the federal government. "At least in part" means that state and local governments, as well as private insurers and employers, may play key
roles in financing and administering such a national program. In
Canada, for example, the so-called national health insurance program is really a collection of ten provincial health insurance programs, regulated and partly funded by the national government.[1]
Recent American proposals to mandate employer-provided health
insurance might also lead to what is here called national health
insurance, so long as the federal government assured both equity
by regulating the coverage of the employer plans and universality
by arranging for health insurance for the unemployed and their
dependents.

From: *Biomedical Ethics Reviews • 1990*
Eds.: J. Humber & R. Almeder ©1991 The Humana Press Inc., Clifton, NJ

The point is that national health insurance does not mean just one thing. Alternative models abound worldwide. The core of what it must mean, the essence of national health insurance, is the use of the authority of the national government to arrange for equitable and universal health insurance. For ease of reference, the simplest model of national health insurance will often be used here: a program in which the federal government establishes a program like Social Security to insure all Americans and finances it through progressive taxation. It is important to remember, though, that this is the simplest model of national health insurance, not the only one. For many practical political reasons it may not be the model best suited to the US at this time. Nevertheless, the moral arguments for national health insurance are largely the same whether the program is administered by the federal or state governments or by private insurers, and whether it is financed exclusively by the federal government or in partnership with states, the business community, and individual Americans.

National health insurance must be contrasted to two alternative arrangements: socialized medicine and free market distribution of health insurance. Under socialized medicine, as in Great Britain, the national government is the owner of most health care facilities and the institutional provider of most care. The national government owns hospitals and employs physicians. In contrast, national health insurance does not give the government an ownership role. Instead, the government functions as a third-party payer, enabling its citizens to pay for care provided by other institutions and individuals. National health insurance is not socialized medicine; it is socialized health insurance.

A free market approach keeps the government out of health insurance, except to tax and to regulate against fraud, abuse, anticompetitive practices, and so on. Health insurance arises out of the interactions of supply—private insurance companies—and demand—employers and individuals. Considerations of access and equity would be secondary to the pressure of market forces. As with all marketplace models, patterns of distribution and the quality of products and services distributed will be shaped by ability and willingness to pay.

National health insurance can be seen as a middle ground between these two extremes of government ownership of health care resources and government exclusion from a health insurance marketplace. On this broad middle stand nearly all of the industrialized democracies of the world. In the West, only Great Britain and, to some extent, Italy have chosen to socialize medicine. The US stands alone among the industrialized democracies for having chosen a largely marketplace model for health care and health insurance.[2]

The Problem

Today in the US, which has some of the best health care in the world, thirty-seven million Americans, or 15.5% of the nation's civilian population, have no private or public health insurance.[3] These Americans do not qualify for Medicaid or Medicare, do not have insurance where they work, and carry no private health insurance. They are quite literally unprotected against the financial costs of health care.

The problem is worsening. The numbers of Americans without health insurance increased by 25% during the 1980s. The problem is especially acute in some sectors of our population. Nearly a third of all Hispanic Americans, the fastest growing and soon-to-be largest minority, are uninsured. Fewer then half of all African-Americans are covered under employment related insurance. One-third of all Americans aged 19 to 24 have no health insurance.

The vast majority (88%) of these uninsured Americans are working or are dependents in families headed by someone who is working. Half of the thirty-seven million live in families where the head of the household is steadily employed on a full-time basis. Eighteen percent are in families headed by a seasonal or part-time worker, and only twelve percent are in nonworking families. Nearly sixty percent of the uninsured live in families with children, and one-fourth of this number live in families with children headed by a single parent.

In fourteen states (Florida, Kentucky, Tennessee, Alabama, Mississippi, Arkansas, Louisiana, Oklahoma, Texas, Idaho, New Mexico, Arizona, California, and Alaska), and the District of Columbia, more than 20% of the nonelderly population are uninsured. In three of those states (Mississippi, Texas, and New Mexico) more than one-fourth of the nonelderly population is uninsured.

Being uninsured and being poor are also closely associated. Some thirty-two million Americans, over thirteen percent of our population, live in households in which the annual income is below federal poverty level (in 1986, the federal poverty level was set at $11,203 annually for a family of four). About one-third of the uninsured have family incomes below the federal poverty level, about half have family incomes of less then $15,000, and three-fourths have family incomes of less then $30,000. The Medicaid program, which was designed to provide access to care for the nation's poor, fails to provide care for more than 10.9 million Americans with annual incomes below the federal poverty level.

Some Results

What are the consequences of these large numbers of Americans having no health insurance? For those who are uninsured, money can be a direct barrier to health care. Despite suffering from higher rates of ill health in general, the uninsured make fewer physician visits and are hospitalized less than those who are insured. In 1987, the Robert Wood Johnson Foundation reported that one million Americans had tried to obtain health care but were turned away due to financial reasons.[4]

The uninsured are also less likely to use preventive services, and especially less likely to receive adequate prenatal care. There is clear evidence supporting an association between inadequate prenatal care and adverse neonatal outcomes—babies who die prematurely or who live without realizing their full potentials.[5] In 1982, 5.5 % of all newborns were uninsured. By 1986, this figure had increased to 8% of all newborns, a 45% jump. Almost 20% of all Hispanic newborns are uninsured. No doubt, lack of insurance contributes to America's embarrassing 19th rank among nations in terms of infant mortality.

Many older children and adult Americans without health insurance simply opt out of the health care system, or they delay seeking care past the point at which care could have been effective. When the uninsured do seek care, it is often in hospital emergency rooms. This is an inappropriate site for most health care and one that can be far more expensive than treating the same condition in a physician's office.[6]

Utilitarian Arguments

According to the principle of utility, we are morally obliged to select those acts or policies that will provide the greatet benefit for the greatest number affected by the act of policy in question.[7] Thus, utilitarianism looks to consequences and is especially attuned to those consequences that bear on the happiness and general welfare of human beings. The theory requires careful attention to balancing consequences so that where there are multiple alternatives for action, the alternative that maximizes good consequences, minimizes bad consequences, or sets the best balance of good to bad consequences is the one that must be chosen.

The realities sketched above suggest that lack of universal health insurance creates considerable amounts of avoidable illness, disability, suffering, and premature death. If national health insurance could minimize these negative consequences, then it is incumbent on us in terms of the utilitarian theory to institutionalize national health insurance.

Health care cannot eliminate all human suffering, and death is inevitable for all of us. But contemporary health care is a powerful means of reducing human suffering and avoiding premature deaths. Thus, the sheer power of contemporary medicine, its ability to bring about good conequences, creates an argument for universal access to it. National health insurance can create universal access, at least to a basic minimum of health care.

In utilitarian reasoning, the relative weight of opposing considerations must also be considered. In this case, there are two. First, would utility be better served by allowing the marketplace to distribute health care? This is generally our approach to the distribution of goods and services in our economy, and in many arenas it works quite well. Moreover, other basics of life—food, clothing, shelter—are distributed largely on a pay-as-you-go marketplace model. Why not health care as well?

The second opposing consideration points to the costs associated with establishing a national health insurance program. Would the benefits of such a program outweigh the costs of increased taxation and increased government bureaucracy?

To the first point, there is a compelling reply. The distribution system at present, one that is largely marketplace based, has simply not worked as measured by the needs of Americans. The fact that other basic needs such as food, clothing, and shelter are distributed on a largely marketplace model cuts both ways. It may mean that a similar argument can be mounted on behalf of universal access to these goods as well. Perhaps it is morally wrong to allow any basic needs to go unmet because of reliance on market distribution.

Moreover, an independent argument can be made for the special importance of access to health care among other basic needs. Although the need for food, clothing, and shelter are universal and pressing, health care needs are different. They are nonstandard and unpredictable in individual cases. When one is hungry, thirsty, or in need of shelter, literally any food, drink, or shelter will suffice. But when one is in pain, has an infection, or a malignancy, a very specific response is called for. Also, needs for food, clothing, and shelter are relatively constant, and therefore, highly predictable. By contrast, none of us is sure when and if we may need access to a hospital emergency room, have a child who needs psychiatric care, or a parent who requires long-term nursing care.

Thus, the answer to the first challenge is that perhaps there should be universal access to whatever is needed to satisfy all basic needs. But even if that argument cannot be made compelling, health care stands out among basic needs for its nonstandard and unpredictable nature.

To the second challenge, there are several responses. First, there are a number of plans that have been put forward that would guarantee universal coverage at little or no increased cost from what is presently being spent on health care.[8] Furthermore, other nations that have adopted national health insurance programs (Canada, for example) spend less per capita and less of their total Gross National Product on health care than does the partly uninsured US.

In spite of the existence of plans that allege little or no increase in costs in a move towards national health insurance, it is reasonable to speculate that establishing such a program will, at some point, mean an increase in taxation. This must be counted as a cost, but so long as the cost is not astronomical, it can be justified by appeal to the principle of marginal utility. According to this principle, a dollar is worth more to a person in need than it is to a wealthy individual because of the greater happiness it can bring at the margin. For example, winning $500 in a lottery might be a momentous event for someone living at the poverty line. It might be a pleasant, but far less significant event, though, for someone whose income was over $250,000 per year. Similarly, the loss of $500 might be staggering to someone at the edge of poverty, whereas the loss of $500 might simply be a bad day for a wealthy individual. Thus, dollars and the resources that they can buy mean different things in different contexts. Under national health insurance, money, through taxation or through some other source of revenue, is moved from those who have plenty to those who have little. It provides protection to those least able to afford health care on their own. Thus, dollars move from the wealthy to the poor. This transfer is justified by the principle of marginal utility because it increases the value of the money and the happiness it can bring. If it cost the average tax payer $500 per year to fund a national health insurance program, it hurts that tax payer less than that "same" $500 helps those who would otherwise be uninsured. Thus, redistribution through the provision or assurance of social services is typically justified by the law of marginal utility.

Concern about the growth of government bureaucracy is certainly legitimate. Bureaucracies tend to become inflexible, self-sustaining, and impersonal. They can frustrate individuality. Generally, therefore, bureaucracies ought not be developed beyond what is necessary. Alternatively put, problems should be dealt with at the least complex level at which they can be effectively addressed—by individual and families first, then by voluntary organizations at the local level, then by national voluntary associations, then by local governments, state governments, and only finally by the federal government.[9] Nevertheless, some issues must be dealt with by the federal government in order to be handled fairly and effectively. Health insurance, because of the national scope of the problem of the uninsured and because of the very basic character of health care needs, is one of those issues that now must be addressed by the federal government. When the government must act, it must seek to do so with the least amount of bureaucracy necessary and with the widest possible range of flexibility and sensitivity to individual and local variations. There are multiple models for how this can be done. Each state might be primarily responsible for its own program, receiving funding and general regulation from the federal government, as in Canada. Or the marketplace might be used to invigorate the program and keep it from petrifying. For example, West Germany has developed such a quasi-marketplace within its national health insurance scheme. Thus, the objection of increased government bureaucracy is an important one, but not a definitive one.

Two other utilitarian arguments in favor of national health insurance are available. In addition to the direct benefit to individuals and their loved ones from universal access to basic health care, there are other benefits available to the entire society, both tangible and intangible.

Universal access to health care will inevitably mean a healthier population. A healthier population is likely to be more productive, to have fewer bed disability days, and to have fewer chronic illnesses. This can mean a great deal to a national economy in an era of increasing international competition and renewed emphasis on productivity. A similar point can be made about national defense and about American culture in general. The healthier a nation, the larger the pool of individuals that can be called on to assist in the national defense. In fact, Great Britain adopted a national health insurance program in 1914 when it became plain that the large numbers of Britons who were unfit for military service constituted a threat to the national interest.[10] A healthy population is also one that is able to contribute more fully to all of the cultural vitality of a nation—to its art, literature, and entertainment. Other things being equal, a healthier population is a more creative population.

Intangibles constitute a second utilitarian consideration. The picture that emerges from statistics on lack of insurance and on health outcomes is not only a negative one, but it is also a potentially divisive one for Americans. Lack of insurance is concentrated among those who are already least advantaged in our society, i.e., the poor and members of minorities.[11] Their health statistics are also uniformly worse than are those of affluent majority Americans. Thus, the specter that this picture evokes is that of two Americas—the haves and the have-nots. Commitment to a national health insurance program would be symbolically significant in as much as it would assert a single national community of concern on health care issues. It would be a sign that Hispanic families are as important as any other American families, that African-American infants are as important as any other infant. It is hard to measure the outcomes of such a symbolic statement, but it seems obvious that it can support better outcomes than an American community increasingly divided over health care.

The Need for Limits

If a utilitarian case for creation of a national health insurance is persuasive, the next obvious issue relates to limits. How much should a national health insurance program cover? What are American's basic health care needs? Plainly, any such program must have limitations since health care can make endless demands on the economy in general and in many specific cases. A useful way of tackling this sort of problem at the theoretical level has been developed by Alan Gibbard.[12] Imagine a person placed behind a hypothetical "veil of ignorance" so that he or she is unable to know the actual conditions of his or her life. This individual is denied knowledge about gender, race, general state of health, and particular health care needs. Now imagine that this individual has a decent minimal income, say 200% of the federal poverty level. Out of this money, the individual must purchase health insurance coverage for his or her entire life. Money not spent for health insurance will be used to cover other basic needs and to provide discretionary income. Thus, this hypothetical individual has an interest in getting the most comprehensive package of health insurance at the lowest possible price.

He or she will also be interested in covering those conditions that are statistically most threatening, but not those that he or she is least likely to suffer from. Thus, this hypothetical individual acts like a utilitarian trying to maximize the impact of limited resources in the purchase of health insurance.

What kind of health insurance would such a hypothetical individual buy? Probably, he or she would buy insurance that would cover adequate prenatal care since the benefits of protection at this early stage in life are so great and its preventive impact so well etablished. Similarly, he or she would want insurance protection to assure adequate pediatric care so as to be protected against a wide range of threats that might lead to premature death. No doubt, he or she would also want insurance for emergency care because of the life and death difference it can make. A basic package of primary care would also be chosen, since this sets the stage for early detection and appropriate entry into all other levels of the health care system. Probably, he or she will also want a package of catastrophic protection in the event that an expensive illness should occur. He or she may also want insurance to provide some protection against the costs of long-term care should it be needed in old age or because of a serious injury.

It is unlikely that this individual would buy insurance protection that would pay for extended care in a persistent vegetative state or care to cover heroic intervention for premature infants with poor prognosis. It is doubtful that he or she would buy insurance against rare diseases or for organ transplantation or other expensive technological interventions.

In short, this hypothetically-placed person would want to buy the insurance package that gives the best chance of leading a long life lived well. He or she would have to use statistical information to arrive at the kind of health insurance that would most likely produce this result. Since this individual is any one of us, so long as we are rational and unbiased, this thought experiment provides the basis for what should be included in a national health insurance program. Coverage that will allow the greatest number of people to lead long lives lived well would be part of what a utilitarian would be obliged to include in a national health insurance program.

Contractarian Considerations

A separate set of moral arguments can be brought to bear on the issue by elaborating on the device of a veil of ignorance to illustrate the central thrust of social contract reasoning. According to the social contract tradition, fairness demands that as we draw benefits from living together in an organized society, so we owe obligations to one another. The moral dimension of life in a political community is founded on a quid pro quo that constitutes an unwritten social contract. This tradition in the West is as old as the argument of Plato's *Crito,* but it has recently been developed by John Rawls.[13]

On the contractarian account, our moral obligations are defined by the character of the social contract, by what we have promised to one another. But there is no actual contract, no record of an explicit promise that we have made to one another. Therefore, we must consider what implied promises we have made to one another by virtue of living together in a society.

One way of articulating which promises we have made implicitly and, therefore, what fairness requires of us, is to consider a group of individuals placed behind a veil of ignorance and set with the task of constructing the basic outlines for a society they are about to enter. The veil of ignorance denies these individuals knowledge about themselves including their sex, race, age, degree of health, and so on. They are allowed to know the general characteristics of human nature, including the facts that we differ in talents and dispositions and that sometimes incentives are effective in calling forth talents and shaping dispositions. Behind this veil of ignorance, these individuals are required to reach unanimous agreement on the social arrangements under which they must live.

In this hypothetical situation, one negotiating strategy appears obvious. Each of these contractors will seek an agreement that gives him or her the best chance of leading a minimally decent life even if his or her real life in society turns out to be the worst one possible. Each contractor will seek to make the worst possible future life the best that it can be, since this worst possible life may be the one that he or she actually leads.

What will they agree to as the basics of a social contract with respect to health insurance? Contractors are unable to predict in detail what their health care needs will be, but they know that people are subject to disease and premature death, that health care can make a difference, and that access to health care might be conditioned by ability to pay or the existence of health insurance. Each will have to consider what life would be like if he or she suffers from a disease in a society that has effective health care, but without the individual financial ability to access that health care. In short, each will have to consider what life would be like without health insurance. Moreover, each one will have to consider what it would be like to be the head of a household in which there are dependents who are also uninsured.

Considerations such as these will lead to unanimous agreement on the need to assure universal access to a minimally decent amount of health care. Some contractors may suggest equal health care for all as the best strategy for dealing with the uncertainties of their future lives. However, this suggestion would not lead to unanimous agreement since contractors will also realize that it might be important to allow for some inequalities in society as a means of increasing the overall well-being of even the worst-off individual. Knowing that humans differ in talents and that incentives are sometimes useful, they might want to allow some individuals to make more money so that their talents will be used effectively in creating wealth that benefits everyone. This would probably mean allowing some individuals to buy more health care or a higher quality of health care than is the norm for everyone else.

In spite of this exception, contractors would insist on access to a minimally decent amount of health care for everyone. A national health insurance program would be the least liberty-restricting method for achieving that result. Socialized medicine might allow for the same result, but would do so in a way that limits more people's liberty. Since it can be presumed that these contractors are also interested in the widest range of liberty possible, they would want to elect a universal program of health insurance that is consistent with the greatest amount of liberty in society. Hence, they will unanimously agree on a national health insurance program.

The moral relevance of this abstract scenario is that the veil of ignorance is a hypothetical device for removing prejudices. Most Americans have health insurance. Most Americans are confident that they can care for themselves and their dependents should illness strike. Therefore, it is not in the direct interest of most Americans to adopt a national health insurance program since they are faring relatively well under the present situation. But to reach this tacit conclusion, most Americans must be sure that they are not about to fall into the categories of other Americans who are uninsured and who have unacceptably poor health statistics. Because race and ethnicity are highly correlated with lack of insurance and poor health outcome, most majority Americans have reason to believe that they will never face these problems. Moreover, most well-educated affluent Americans also have reason to believe that they will fare well in a competitive society, and therefore, not fall into poverty.

The veil of ignorance acts to strip away this confident self-knowledge and to make us each face the moral question, "What would I consider a fair arrangement of society if I could be any member of that society?" Looked at this way, it is morally unacceptable to allow vast numbers of Americans to go without guaranteed access to a decent amount of health care. No one of us would agree to that if we knew that it were possible to be among the least advantaged. It simply wouldn't be acceptable as a fair social contract. If fairness from a position free of prejudice demands universal access to basic health care, and we are morally bound to be both fair and nonprejudiced, then we are morally bound to guarantee universal access. National health insurance would guarantee it.

What would be the details of the national health insurance program that prejudice-free contractors would agree to? This is hard to answer in specific terms, but generally, contractors would be more concerned for the program's impact on each individual than they would be for achieving the greatest good for the greatest number. In other words, they will not use utilitarian reasoning. Instead, in deciding what to cover, they will have to consider what would be the situation if they were among those with certain kinds of diseases or disabilities. Consequently, a contractarian-based national health insurance program would cover more than what would lead to a long life lived well for the greatest number of individuals. The program would have to be structured and services included on the basis of what would make the least life among us the best it can be. This would be a highly demanding standard, but on contractarian accounts, it is a standard demanded by considerations of fairness. It is what each person would choose if he or she were behind a veil of ignorance.

Egalitarian Considerations

A third kind of argument in favor of national health insurance can be developed if the nature of what it is to be a human person is articulated. On one persuasive account, to be a person is to be fundamentally different from a thing.[14] The value of a thing is relative and is established, at least in part, by extrinsic considerations. For example, the value of an automobile is determined by its price in the marketplace. Its price will increase when demand increases or supply decreases. Since these changes occur outside of the car, i.e., when nothing about the car itself has changed, price is an extrinsic measure of value.

In contrast to the price of an object, a person has dignity, a measure of value that is both absolute and intrinsic. "Absolute" means that no relative estimation can be made of the value of a single human life. In numerical terms, the value of each human life might be said to be infinite, literally incalculable. This absolute estimation is grounded in the fact that the value of a human life is "intrinsic," i.e., internal to what it is to be a human being. Human beings do not derive this fundamental value by virtue of any achievement, strength, beauty, or special talent, nor do individuals lose it by virtue of any flaw or shortcoming. This value is inherent in each individual simply by virtue of being a human person. So while things have a price that establishes their relative value with respect to conditions outside themselves, human persons are priceless. The value of each is absolute and intrinsic.

Because the value of each human life is, thus, incalculably great, all persons must be considered equals in terms of human dignity. Clearly, there are empirical senses in which people are unequal. Some are stronger than others. Some are smarter. Some are more beautiful. Some have special abilities. But since none of these differences relate to the source of dignity itself, they do not affect the main point that humans are moral equals. This consideration forms the foundation of our political commitment to equality before the law and to the recognition of universal human rights. Thus, reflection on the nature of the human person gives rise to an egalitarian perspective, animated by the view that, on the most important grounds, human beings are fundamentally equal.

A significant parting of the ways among egalitarians occurs when this insight is applied to social policy. Some egalitarians insist on substantive equality, i.e., achieving equality of results or outcomes. On this interpretation of egalitarianism, the dignity of each human person creates a social obligation to make people as empirically equal as possible. This has the ironic but inevitable implication that in order to make people equal, we must treat them unequally. To make people equal in income, for example, would require giving to those who have little, taking from those who have much, or both.

Other egalitarians insist on procedural or formal equality. On this account, human dignity requires a commitment to treating people equally, to creating systems of due process so that each of us is accorded the same opportunities from our basic institutions. This also has the ironic and inevitable implication that equal treatment in a situation of empirical inequality leads to inequality in outcome.

The egalitarian argument for national health insurance can be constructed using both of these insights at different points of the argument. The motivation for the establishment of national health insurance is plainly substantive egalitarianism. Because of the equal human dignity of all persons, an egalitarian would want to make health care universally available so as to overcome or lessen inequalities in health outcome, so that, for example, infant mortality among African-Americans could be brought closer to the white rate (at present, it is twice as high). Thus, substantive egalitarianism provides the justification for national health insurance.

However, a national health insurance program would have to be structured on procedural egalitarian lines. All Americans would have to receive the same basic package of protection. Each would be covered for the same schedule of health care services, and each would have the same deductibles and limitations. Obviously, people will access the health care system for different reasons, and thus rely on the national health insurance protection in different ways, but every citizen will be brought through the process as an equal. Thus, the structure and the operation of the system would embody procedural egalitarianism.

Seen in this way (as a compromise between substantive and procedural egalitarianism), national health insurance again holds the middle ground between socialized medicine and the marketplace model. Insistence on substantive equality exclusively might create a case for socialized medicine. Only if the government becomes fully involved in health care production and distribution would it be able to finely modulate health outcomes so as to guarantee greater equality of results. On the other hand, the marketplace model represents an ideal procedural equality. Every person is treated the same inasmuch as each has the same liberties to enter or leave the marketplace, to produce or to consume, and to agree or refuse to trade. These extremes have their characteristic flaws. Socialized medicine risks creating an ossified bureaucracy and the moral difficulties of unequal treatment in order to bring about more equal results. The marketplace has largely failed the uninsured, and through its commercial equality of process, has generated large-scale inequalities in outcomes. National health insurance avoids both extremes.

There are two further considerations that lead to egalitarian support for national health insurance. First, all of our other equal human rights, the rights that are articulated in the US *Bill of Rights,* require life and a degree of health to be exercised and enjoyed. Thus, our whole structure of equality before the law and of equal standing in the political process relies on our prior ability to live and to thrive. Therefore, a right to health care guaranteed through universal insurance provides the basis on which all of our other rights can be realized and enjoyed.

The second argument turns on inequality in our society as a whole. From a substantive egalitarian point of view, the wide range of inequalities in our society is unjust. That thirteen to fifteen percent of our population live in households below the federal poverty level, for example, is in itself wrong. Establishment of a national health insurance plan will not directly address such stark inequalities, but it will help. There is a close association between illness and poverty, though the causal relationship is much debated. In many cases, poverty produces illness and injury. In other cases, illness and injury bring on unemployment and poverty.

Assuring access to health care will help to break the link between disease and poverty, and therefore, help soften the impact of the economic inequalities of American life.

Libertarian Objections

If the moral case for national health insurance can be made using arguments from the utilitarian, contractarian, and egalitarian traditions, where does the conceptual resistance to the idea come from? Obviously, there are many practical and political problems that stand in the way of constructing a national health insurance program. But is there a moral objection to it?

The main moral objection to national health insurance would come from the libertarian tradition. According to this theory, the single most important moral value is individual liberty.[15] Government action of any sort is restricted in principle by the side constraint of deference to individual liberty. No other considerations—not happiness, fairness, or equality—can justify government infringement of individual liberty. Put in other terms, the negative right of noninterference with individual liberty consistently trumps claims for any positive rights of access to basic goods and services.

This conceptual point can be made into two specific objections to national health insurance. First any national health insurance program will have the inevitable implication of restricting the liberty of individuals to buy and sell health insurance on the open market. In the simplest plan, the government would be the sole provider of health insurance and would ban all private sales of insurance packages that compete with the national program's coverage. Even on more pluralistic models where private health insurance companies act as brokers for the national health insurance program, their liberty to freely enter and leave the marketplace would be considerably limited. Thus, the first objection is that the individual liberty to enter into contracts would be wrong-fully restricted. Regardless of the nature of the goal at hand, this is an unacceptable infringement on individual liberty.

The second libertarian objection is more general in nature. If the government were to establish a health insurance program, in all likelihood, taxes would have to be raised. Even on the sanguine assumption that we would pay no more totally for our health care under a national health insurance program than we pay at present, a national program would probably mean that money we presently pay for insurance in private employment arrangements would be raised through more direct forms of taxation.

But for libertarians, taxation is problematic. The potential implications for liberty are obvious when the consequences of failure to pay taxes are considered. The government will force an investigation and if it is not satisfied, an individual citizen may face imprisonment. Thus, there is a direct connection between taxation and liberty, and taxation for purposes beyond a legitimately limited government is wrongful coercion. Since taxation implies a threat to individual liberty, the power of the government to tax must be kept to its smallest legitimate expression. This means that a government can impose taxes only to prevent and correct infringements on individual liberty. It is illegitimate for the government to impose taxes for other purposes, including efforts to promote access to basic goods and services.

There are persuasive responses to these libertarian objections. First, there is no inherent right to buy and sell all and any goods and services. A democratic community has an obligation grounded in its responsibility to the common good to regulate the marketplace, including the prohibition of sales of certain kinds of goods and services. Dangerous products, fraudulent services, or the exchange of goods and services that are inherently destructive of a community's morally valid purposes can properly be prohibited. Thus, the view that the insurance industry has an inherent moral right to sell their product is an overstatement that cannot be accepted.

There is a compelling practical issue at stake here. The attempt to spread the benefits of health insurance by the use of the free market has failed and is creating a public health crisis. The numbers of uninsured are climbing. An historical perspective is useful in seeing why this is so.[16] Health insurance in America began by using community rating, i.e., by establishing a premium and a package of coverage based on the average experience with disease and injury of individuals across a whole community. But marketplace competition has led to the demise of community rating. It has been replaced by experience rating, the basing of premiums and coverage on the actual experience of insured groups or individuals. Insurance companies seek to offer lower rates and better coverage to groups and individuals with more favorable experience. The shift from community rating to experience rating also means higher process or denial of coverage for those people who are most in need of insurance: those who have had significant health problems already and those who are in high risk groups to have health problems. Thus, the marketplace operating under its own logic has increasingly worked against the social justification of insurance, i.e., against the pooling of risks. The normal mechanisms of the marketplace are forcing insurance companies to fine-tune their underwriting so as to cover only those who are least likely to use the insurance that they are buying. A national health insurance program would reestablish the widest possible pool. It would thus state in the clearest moral terms that the nation is a community of persons that share each other's health-related risks.

Libertarian moral objections against taxation simply cannot be accepted as a matter of principle. Certainly, members of a democratic community can disagree on prudential grounds about different kinds of social programs and different rates of taxation. Clearly, there can be times when people are overtaxed, but the libertarian position is not the practical one that present rates of taxation are too high. Rather, it is the theoretical position that it is morally wrong to collect taxes for purposes that go beyond the most narrow scope of government. This extreme view has never been accepted in our national experience. In the US and in democratic communities worldwide, it has been accepted as a matter of political common sense that the government has an obligation to cultivate the common good and to improve the quality of life for individuals. This mandate is properly constrained by the obligation to respect individual human rights, but the degree to which libertarians would constrain government is unacceptable. There can be no human right to be free in principle of taxation whose purpose is to cultivate the common good. Again, there is plenty of room for significant moral and political differences about how far the government should go in cultivating the common good and what is an appropriate level of taxation. Reasonable people of good will can disagree on these matters. It is simply an unacceptable extremism, though, to rule out the use of taxation for all public purposes beyond the protection of individual liberty.

In general terms, the problem with libertarianism lies in its totalizing character. Certainly, freedom is an important human value, but libertarianism makes freedom the only human value, and thus, is the most extreme form of individualism. As such, it is inimical to the interest of any community. Regardless of one's social philosophy or ethical theory, there is an inherent tension between the individual and the community. Finding the proper balance is never an easy task, but a totalizing concept of individual freedom is surely an inappropriate imbalance. There simply is no way for individuals in a democratic community to reform and improve their common life together if the first moral assumption is that the single most important value in every political context is individual liberty. Democratic communities must have the rightful authority to address problems of significance to the common good by rejecting some claims for liberty and imposing reasonable taxation.

In conclusion, libertarian objections to national health insurance constitute an important warning against allowing the government to become an oppressive force and against imprudent spending. It is a useful reminder of the importance of individual liberty. But as a moral objection against national health insurance, it cannot be accepted. Its objection against national health insurance is the same objection that it would have to level against Medicare and Medicaid. It would have to have the same objection against federal aid to children in poor families, the food stamp program, support for higher education, and all public support for health care, education, housing, and so on. This is simply too totalizing a concept of individual liberty.

Other Practical Objections

There are several other objections to national health insurance that must be touched on. They are not objections at the level of moral principles, but they are important practical objections that may have significant moral dimensions, especially if they anticipate seriously negative consequences. There are four major practical objections.

1. National health insurance is inimical to the practice of good medicine. It would interfere with and destroy trust in the doctor/patient relationship. It would diminish the autonomy of doctors and reduce their ability to make a just livelihood. It would make medical practice so unappealing that doctors will leave practice in the US in large numbers and young people will no longer be attracted into the profession.
2. Research and development of new medical technologies would inevitably be sacrificed by a national health insurance program. There would be less money for basic research, and therefore, less progress in the battles against disease and premature death.
3. National health insurance would create a bureaucratic nightmare. It would be impossible to administer efficiently. It is simply a task that government cannot do well.

4. A national health insurance program would ultimately prove to be far too costly for the American economy. To commit ourselves as a nation to provide financing for health care for all Americans would lead to bankruptcy. Health care needs can constitute a "bottomles pit," since no one wants to die and our abilities to intervene against disease and death are exceptionally expensive.

These are important objections. Each deserves a detailed rebuttal (a project that cannot be accomplished here), although a sketch of a response can be provided for each objection.

1. A national health insurance program need not have an adverse effect on the practice of medicine. National health insurance would not socialize the practice of medicine, but would socialize the financing of health care. National health insurance would provide patients with the financial wherewithal to seek care from physicians and other health care providers.

   National health insurance need not have any negative impact on the doctor/patient relationship. Indeed, it may enhance the doctor/patient relationship by removing financial pressures that are presently building on both patients and doctors. The present myriad of DRGs, PPOs, capitation plans, RBRVSs, commercial ventures by doctors, increasing deductibles, copayments, caps, and so on are beginning to take their toll on the doctor/patient relationship. Many are concerned about conflicts of interests on the doctors' side and patients are entering relationships more skeptically. National health insurance holds out the promise of preventing the further deterioration of the doctor/patient relationship by creating a clear and consistent stream of financing that can be made equitable to both parties.

   Payment to physicians need not be unfairly low. Experience in other nations indicates that national health insurance has actually helped many physicians since they no longer have the burden of charity care nor bad debt; all their patients are paying patients.[17] It is likely, however, that some physicians who are making incomes at the very highest range may not see those incomes supported by a national health insurance program. But most Americans will not see this as unfair.

Will doctors leave the system in droves? Will young people refuse to join the medical profession? No one can predict the future with certainty, but the experience of other nations indicates that although some physicians may become dissatisfied with this kind of change in the financial environment, it will not have a negative impact on large numbers of medical professionals. There was no exodus of Canadian doctors into the US when Canada adopted its national health insurance program in the 1960s. Doctors in Western Europe, Japan, and Australia are not leaving their practices because of national health insurance programs in their nations. Thus, international experience suggests that medicine will remain an attractive career for many and that doctors and other health care professionals can live and earn a just wage under a national health insurance program. On the other hand, there is a rising tide of discontent among doctors with the present American system.[18]

2. Research and development may or may not suffer under a national health insurance program, depending on the level of funding provided for it. Public funding might continue unchanged, irrespective of the establishment of the national health insurance program. For example, the National Institutes of Health would probably remain independent of a national health insurance program, its research funding unaffected by the existence of an insurance program that pays for health care services. The practical question is to what degree is society willing to fund continued research and development in health care. If the public

sees this as a valuable investment, other things being equal, then the funding will be adequate. If the public begins to feel that research and development is not worth the investment, then it will likely dwindle. This is generally the case, though, irrespective of a national health insurance program.

3. Would national health insurance create an administrative nightmare? Many commentators believe that health care in the US is an administrative nightmare at present.[19] In the Canadian national health insurance program, there is one-third party payer, and therefore, one set of forms and one set of administrative procedures. By contrast, in the US, there are over 1500 payers at every level of government and throughout the private health insurance industry.[20] Each one of these payers has its own forms and administrative processes. Which is likelier an administrative nightmare?

In spite of claims about efficiency, it is not always true that the private sector does a more efficient administrative job than does government. The Medicare program, for example, operates with a smaller administrative overhead than does the average private health insurer.[21] Additionally, the government takes no margin or profit, creating additional efficiencies in the system. Again, the international experience here is that a national health insurance program can be administered with a degree of efficiency that is acceptable to the general public. Every system will have its flaws and every system will have its inefficiencies. The question is whether or not a proposal for national health insurance will have inefficiencies and administrative complexities that are intolerable. The answer seems to be no.

4. Would a national health insurance program ultimately prove too costly? This is an exceptionally important question because guaranteeing universal access to basic health

care would certainly increase costs. Guaranteeing unlimited universal access would be an impossible goal. Therefore, reason demands that any national health insurance program do explicitly what is now being done implicitly, i.e., the program must ration. Rationing is accomplished now by the marketplace, by people's insurance status and ability to pay, by limitations set by private insurers and Medicare and Medicaid, and by the general limitations implied by our health care institutions and resources. But establishment of a national health insurance program would force us to state quite plainly as a nation the limits beyond which we cannot be asked go.

Setting limits will be a difficult process, but it is a necessary process. Presently, many limits are being set in corporate board rooms without any public debate, without any inspection of the moral arguments for limitations, and without any consensus in a democratic community. A national health insurance program would bring the debate into the public realm and invite the national community to arrive at a consensus on how much health care is an American's right as we approach the twenty-first century.

Attitude, Feelings, and Character

The moral case on behalf of national health insurance has now been made. Utilitarian, contractarian, and egalitarian arguments have been used on its behalf. Libertarian objections have been answered, as have certain practical objections.

But there is another dimension to morality that must be articulated. Morality is a guide for shaping conduct. In the practical arenas of life, conduct is shaped not by abstract argument of the sort used here, but by moral attitudes, feelings, and traits of character. This is especially true of choices about social policy. It is important to articulate the moral arguments for social policy, but when all is said and done, Americans will choose their social direction on the basis of their attitudes, feelings, and traits of character.

From this perspective, the moral case for national health insurance really turns on how we think about ourselves as a people, how we feel about the sufferings of our fellow citizens, and whether we have the strength of political will to do what is required by our attitude and feelings.

It is possible that Americans will continue to think of themselves primarily as individuals in diverse communities who join together only in wars or in times of national trauma or jubilation. We may continue to think of ourselves as radically individual and separate one from another. We may continue to allow deep-seated racism and ethnic and religious divisions to influence the crafting of social policy. On the other hand, the debate about national health insurance is an invitation for us to see ourselves as one national community. It is an opportunity to pool our risks and to consider our fates bound together. It is a moment in history when we can build social solidarity and overcome the gulf that threatens to create two Americas.

These points about attitude might be said about many contemporary national struggles—illiteracy, homelessness, drug abuse, and so on. But the issue of health care has a special affective poignancy. We are quite literally talking about life and death, about the quality of life, and about human suffering. Americans without health insurance face needless physical pain, psychological distress, and the existential frustration of lives that are shortened or lived with less quality than they might have. The debate about national health insurance is a challenge to feel with these Americans. It is a challenge to redirect some of our natural moral feelings of sympathy and concern for known individuals into a social program that will save unknown but statistically significant numbers of our fellow citizens. And it could be a means of protecting those feelings of sympathy. The news media is filled constantly with images of suffering, violence, and shortened lives. This daily drumbeat can coarsen and attenuate our feelings for others. It can make us jaded. Institutionalizing a national health insurance program is a way to give positive vent to appropriate moral feelings and to shelter them from the constant assault of pictures and accounts of human suffering and premature death. National health insurance is a way of doing something about this panorama of suffering and, thus, maintaining civilized sensibilities.

Finally, the moral case for national health insurance turns ultimately on strength of character in a social setting. For an individual, strength of character is the opposite of weakness of the will. It is the ability to move consistently from the knowledge of what is right to doing what is right. It is the ability to build habits and dispositions that will guide action spontaneously in accord with reasons that are compelling on reflection.

Weakness of will in a society is the inability to translate ideas and principles into practice, into institutional form. Our historical national commitments to building prosperous communities, shouldering our fair burdens, and struggling for equality all point in the direction of establishing a national health insurance program. Too often, the libertarian objection is not really a moral argument, but is instead an expression of weakness of will. It can be a conceptual cloak for selfishness, for a desire to pay less in taxes, or for the comfort of not having to identify with the sufferings of other Americans. It can be an excuse for dodging the burdens of doing what, at some other level, is known to be right. From this perspective, the national health insurance issue is a test of political will, a test to see if Americans have the will to create in reality the decent society so cherished in theory. It is an opportunity to join the rest of the industrialized democracies in eliminating the suffering, unfairness, and indignity of medical indigency. In the final analysis, the moral case for national health insurance rests on an appeal to America's character as a nation.

Notes and References

[1]Theodore Marmor, Wayne Hoffman, and Thomas Heagy (1983), "National Health Insurance: Some Lessons from the Canadian Experience," *Political Analysis and American Medical Care* (Theodore Marmor, ed.), Cambridge University Press, Cambridge, UK, pp. 165–186; also on Canadian health care system, *see* John K. Iglehart (1989), "The United States Looks at Canadian Health Care," *New England Journal of Medicine*, **321**, 1767–1772 .

[2]Sr. Amata Miller (1985), "The Economic Realities of Universal Access to Health Care," in *Justice and Health Care* (Margaret Kelly, ed.), Catholic Health Association, St. Louis, MO, pp. 109–130.

[3]The data in the following paragraphs comes from Charles J. Dougherty (1988), *American Health Care: Realities, Rights and Reforms*, Oxford University Press, New York, NY, ch 1; Patricia A. Butler (1988), *Too Poor To Be Sick*, American Public Health Assoc., Washington, DC, pp. 1–26; Robert Blendon (1988), "What Should Be Done About the Uninsured Poor?," *JAMA*, **260**, 3176–3177; and "37 Million Americans Found Lacking Health Insurance," *American Medical News*, May 19, 1989.

[4]H. Freeman, R J. Blendon, L. H. Aiken, S. Sudman, C. Mullinix, and C. R. Corey (1987), "Americans Report on Their Access to Health Care," *Health Affairs*, **6**, 6–18.

[5]*Death Before Life: The Tragedy of Infant Mortality*, National Commission to Prevent Infant Mortality, Washington, DC, August 1988; also, Paula Braveman, Geraldine Oliva, Maria Grisham Miller, Randy Reiter, and Susan Egertee (1989), "Adverse Outcomes and Lack of Health Insurance Among Newborns in an Eight-County Area of California, 1982–1986," *New England Journal of Medicine*, **321**, 508–512.

[6]President's Commission for the Study of Ethical Problems in Medicine and Biomedical and Behavioral Research (1983) *Securing Access to Health Care*. Government Printing Office, Washington, DC **1**, 75–76.

[7]The key historical text is John Stuart Mill (1971), *Utilitarianism* (Samuel Gorovitz, ed.), Hackett Publ. Co., Indianapolis, IN.

[8]For example, the "Physicians for a National Health Program Proposal" claims there would be no net increase in costs with their plan for universal coverage. The "Consumer Choice Health Plan for the

1990s" by Enthoven and Kronick claims a negligible impact on the federal budget. For a comparison of these and four other proposals for universal coverage, *see* Rosemary Kern and Jack Bresch (1990), "Systemic Healthcare Reform: Is it Time?," *Health Progress*, January/ February, 32–44.

[9]In Catholic social thought, this is called the principle of "subsidiarity." *See*, e.g., The US Catholic bishops letter (1986), "Economic Justice for All: Catholic Social Teaching and the U.S. Economy," *Origins*, **16**, secs. 98–101.

[10]Paul Starr (1982), *The Social Transformation of American Medicine* Basic Books, New York, NY, p. 239.

[11]*See* note 3, Dougherty, pp. 3–15; Robert Blendon et al. (1989), "Access to Medical Care for Black and White Americans,"*JAMA,* **261,** 278–281.

[12]*See* note 6, Allan Gibbard, "The Prospective Pareto Principle and Equity of Access to Health Care," **2**, 153–178 .

[13]John Rawls (1971), *A Theory of Justice,* Harvard University Press, Cambridge, MA; for application to health care, *see* Ronald Green (1976), "Health Care and Justice in Contract Theory Perspective," in *Ethics and Health Policy* (Robt. Veatch and Roy Branson, ed.), Ballinger, Cambridge, MA, pp. 111–126.

[14]This is taken from Immanuel Kant (1981), *Grounding for the Metaphysics of Morals* (James Ellington, ed.), Hackett, Indianapolis, IN.

[15]This account of libertarianism relies on Robert Nozick (1974), *Anarchy, State, and Utopia,* Basic Books, New York, NY.

[16]*See* note 10, Starr, pp. 328–330.

[17]*See* note 1, Marmor, Hoffman, and Heagy, pp. 174–175.

[18]*See* the three-part series by Lawrence K. Altman with Elizabeth Rosenthal, Lisa Belkin, and Gina Kolata, "Doctors in Distress," *New York Times,* Feb. 18–20, 1990, pp. 1 and ff.

[19]*See*, e.g., David Himmelstein and Steffie Woolhandler (1989), "A National Health Program for the United States," *New England Journal of Medicine,* **320,** 102–107.

[20]Amy Carlsen (1989), "Doctors Urge a National Health Plan, " *HealthWeek,* January 23, p. 15 .

[21]"Administrative costs are also lower in Canada, where overhead and paperwork absorbs about 3% of the health budget. In the United States...the 1500 private insurers have overhead costs of close to 12%,

covering items like marketing, reserves for future claims, taxes, and profits, Federal data shows. The Federal Medicare and Medicaid programs have overhead costs of about 3.5%." Milt Freudenheim, "Debating Canadian Health 'Model'," *New York Times*, June 29, 1989, p. 23.

Can National Health Insurance Solve the Crisis in Health Care?

William B. Irvine

Introduction

America has toyed with the idea of national health insurance for decades. In the mid-1960s, the idea was popular enough for a limited form of national health insurance—viz., Medicare—to be enacted. This was followed by a push for a generalized form of national health insurance. In the late 1970s, a number of national health insurance bills were introduced in Congress.[1] When none of these bills were enacted, the concept of national health insurance fell into a period of decline, and by the early 1980s, the concept had relatively few advocates.

In the last few years, though, all this has changed. The idea of national health insurance has not only been resurrected, but a case can be made that this time, the chances of passage of some form of national health insurance are better than ever.

The principal cause of this resurrection is the crisis in American health care. The crisis in question is not one of quality, for a case can be made that sick Americans have a better chance than ever of recovering. Rather, the crisis involves the cost of providing health care. In the last few decades, medical costs have soared

From: *Biomedical Ethics Reviews · 1990*
Eds.: J. Humber & R. Almeder ©1991 The Humana Press Inc., Clifton, NJ

compared to other costs, and in the last half of the 1980s, the rate at which health care costs increased accelerated sharply.

Corporations offering their employees health insurance have been hit hard by these increases. According to a survey of members of the National Association of Manufacturers, for example, health care costs for employers soared nearly 30% in 1988.[2] Another survey found that the cost per employee of employer-sponsored health care plans rose 18.6% from 1987 to 1988, an increase more than double that of the preceding year.[3] Yet another survey found that outpatient care cost employers 25% more in 1988 than in 1987.[4] Furthermore, there is reason to think that health care costs for corporations will continue to rise in the future.[5]

This surge in the health care costs of corporations has caused a number of those that formerly opposed national health insurance to revise their views on the subject. Bethlehem Steel and Chrysler, for example, recently declared their support of national health insurance.[6] It is striking that "big business," which is so often opposed to government provision of services performed by the private sector (and this is what national health insurance would involve) are uniting to petition the government to provide such services.

Many employers have responded to increased health care costs in part by cutting back on medical coverage and/or by passing on a portion of these costs to their employees. According to the Service Employees' Union, for example, their members' share of health insurance costs surged 70% from 1987 to 1989.[7] This rise in employees' health care costs has become a new source of friction between employers and employees. Indeed, in the telephone company strikes of the summer of 1989, the major issue was not the traditional one of wages or of work conditions, but was instead the proposal to make employees pay a bigger share of the cost of health benefits.[8]

The fact that they are no longer shielded by their employers from increases in the cost of health care has brought the health care crisis home to many American families. They no longer regard it as an issue for insurers and academicians to fret about, but one for their elected officials to deal with. Many polls show popular support for some form of national health insurance, and it is likely that many of its advocates have economic rather than ideological grounds for their advocacy. In particular, they hope that national health insurance can somehow curb soaring health care costs.

Besides those Americans who are being asked to contribute more to health care plans, we should keep in mind the plight of the approximately 37 million Americans who have no health care insurance at all.[9] One reason that this group lacks health care insurance is the high cost of health care, and it is likely that many members of this group look to national health insurance as a form of health insurance that they can afford.

The support of corporations and citizens, of course, constitutes an impressive political force, but support for national health insurance doesn't end here. Even physicians, who in the past staunchly opposed national health insurance, seem to be jumping on the bandwagon. The American Academy of Pediatrics, for example, has proposed a national health insurance plan that would provide coverage to children and pregnant women;[10] Arnold S. Relman, editor of the prestigious *New England Journal of Medicine,* has called for enactment of national health insurance;[11] and it is even rumored that the American Medical Association, long one of the most vociferous opponents of national health insurance, is considering support of such a plan.[12]

In summary, there is an impressive base of political support for national health insurance, and it is altogether conceivable that the support will this time be enough to turn national health insurance from a concept into a reality.

Now, much of the present support for national health insurance is based on the assumption that national health insurance can solve America's health care crisis by making health care more affordable. It is this assumption that will be examined in what follows. The goal is not to evaluate the desirability of national health insurance in general; such a topic would require a book, if not several books. Rather, it is to evaluate the desirability of national health insurance as a solution to the above-described health care crisis. As the reader shall discover, I do not think that national health insurance can succeed in reducing or holding down health care costs. Indeed, I argue that the federal government is best viewed not as part of the answer to the health care crisis, but as a causal factor in the crisis.

Why the Crisis?

Before asking whether national health insurance can solve America's health care crisis, it will be useful to inquire briefly into the causes of the crisis. Until we recognize the root causes of a crisis, we are unlikely to come up with a genuine solution to it, i.e., a solution that deals with its causes rather than merely masking its symptoms.

As is true of any complex economic phenomenon, the health care crisis has a number of causes. Let me describe four contributors to the recent rise in the cost of health care.

1. *Advances in medical technology.* One reason that health care is so expensive is because of advances in medical technology. Generally speaking, new medical treatments are expensive. This means that if we choose to use newly developed medical treatments—and we almost invariably do—we are likely to push up the cost of medical care.

    Stated differently, one reason why medical care was cheaper sixty years ago than it is today is simply that there was less that doctors could do for their patients then. Many of the patients who would have been untreatable in 1930 are treatable today. By way of illustration, consider the fate of a child born with a defective heart. Sixty years ago, this child would simply have died and the costs of medical treatment would, therefore, have been minimal. These days, a variety of treatments can give this child years or even decades of life. These same treatments, however, can easily generate health care bills in the hundreds of thousands of dollars.

I don't mean to suggest that breakthroughs in medical technology always make health care more expensive. There are many breakthroughs that have had the opposite effect. Consider, for example, the discovery of antibiotics. This discovery allowed doctors to treat in their offices and at insignificant cost, patients who sixty years ago would have spent days or weeks in the hospital. My claim is that, for whatever reason, breakthroughs that increase the cost of health care seem to outnumber (and at any rate "outweigh") the breakthroughs that decrease the cost of health care.

2. *Inelastic demand for health care.* A second reason why health care is so expensive is that the demand for it is, to use economists' terminology, inelastic. An example of a good with an elastic demand would (for most people) be lobster. When the price of lobster rises, fewer are purchased; when the price falls, more are bought. On the other hand, health care is the classic example of a service for which there is (for most people) an inelastic demand. If a certain operation is the only thing standing between a person and death, he or she will elect to have it, even though it is very costly and may mean falling hopelessly into debt. In such a case, it matters little whether the operation will cost $100 or $100,000; he or she will elect to have it performed.

The inelasticity in the demand for health care means that steep rises in price will not be met by reductions in demand, at least not for most people. The law of supply and demand, although not defunct, operates with diminished vigor.

3. *The "shield" effect.* A third reason why health care is so expensive is that insurance plans are generally structured so as to shield policy-holders from the full cost of any treatments they may receive. This, of course, is the *raison d'etre* of insurance plans. In shielding policy-holders, though, insurance plans make them relatively indifferent to the cost of the health care they receive, and this in turn makes health care costs more buoyant than they would otherwise be.

    On a personal level, my employer-provided insurance plan (as I understand it, at any rate) makes me pay, each calendar year, the first $200 of health care costs, only 20% of the next $10,000 of health care costs, and none of any health care costs beyond this. In other words, after the first $200 of health care costs, my plan shields me from part or all of the cost of health care that I receive, and in so doing, makes me relatively indifferent to the cost of any treatments I receive.

    Suppose, for example, my doctor told me that I needed a certain operation and that he could perform the operation in question at a certain hospital for $5,000. If I had to pay the full $5,000, I would probably do a fair amount of research before taking my doctor up on the offer: I would undertake some "comparison shopping." I might ask for a second opinion. I might inquire into less expensive treatments. I might shop around for a less expensive surgeon or for one affiliated with a less expensive hospital. The reason I would go to this trouble is simple: I would get to keep 100% of every dollar in reduced costs. If, on the other hand, I have to pay only 20% of the $5,000, I will be less concerned with shopping around for a better deal; after all, I will get to keep only 20 cents of every dollar that

I can save. (Indeed, by shopping around, I will be doing my insurance company a big favor, since it will get to keep 80 cents of every dollar I save.) I would be much more likely to give my doctor the go ahead without further thought. I don't think that my behavior in this regard is unusual.

Lots of Americans are exposed to the above-described shield effect, and because of this, lots of Americans are relatively indifferent to the cost of the medical treatments their doctors prescribe. This, in turn, has a levitating effect on the cost of medical services.

Before ending my discussion of the shield effect, let me point out one bit of irony. Even if Americans overcame the shield effect and became sensitive to the costs of the medical treatments their doctors prescribed, they might find themselves frustrated when it came time to "comparison shop" for lower prices. For as it so happens, in America it is quite difficult to establish, before most operations, what the cost of the operation will be. In America, doctors and hospitals rarely give their patients binding estimates of the sort you might get from an auto body shop. Instead, the parties involved will give you ballpark figures that often have little connection with the bills the patient subsequently receives.

As a result, many patients go into operations with only a vague idea of how much their operation will cost, and for months after the operation, the typical patient will go to his or her mailbox each day to find bills for things that he or she had not realized were involved in the operation and from doctors whom he or she had never heard of. For many patients, anesthetists, histologists, and radiologists are shadowy figures until, a month or two after the operation, the patient obtains tangible evidence of their existence in the form of bills for services rendered.

Not only is it difficult to get a doctor to specify how much a treatment will cost, but even if a person is successful, it is often difficult for him or her to find out ahead of time how much of this bill his or her insurance company will be willing to pay.

Of course, to the extent that it is difficult for health care consumers to comparison shop, these consumers will have that much more reason to be indifferent to the cost of a particular medical treatment.

4. *Medical monopolies*. A fourth and very important reason why health care is so expensive in America is that most of the health care professions enjoy monopoly status. Doctors, nurses, pharmacists, and other health care professionals are protected from much potential competition by a number of state and federal laws.

Consider, for example, the situation of medical doctors. In America, not just anyone can practice medicine; the state allows only those who have been licensed to do so. (Try to practice medicine without a license, and you will find yourself in jail.)

The state could, though, have the policy of licensing anyone who was competent, regardless of how he or she gained this competence. In particular, the state could allow anyone to operate a medical school, and then, by means of an extensive testing process, decide which graduates of these medical schools were competent to practice medicine. In America, however, the state has chosen a different route; besides restricting who can practice medicine, the state also places restrictions on who can teach the practice of medicine. It does this by licensing only those who are graduates of accredited medical schools and who have interned at approved hospitals.

The state also takes steps to make sure that members of the various health care professions do not compete with each other. Thus, nurses are allowed to give shots, but generally aren't allowed to decide which shots to give; pharmacists are allowed to fill prescriptions, but generally aren't allowed to decide which medicines to prescribe; dental hygienists are allowed to clean teeth, but generally aren't allowed to fill cavities; and so forth.[13]

In similar fashion, not just anyone can start a hospital. You typically need permission from the state to do so.

Combine all these restrictions on who can practice medicine and on how it can be practiced, and you have a potent government-created-and-preserved monopoly. Also, by granting the health care professions monopoly status, we protect them from competition from other health care professions and from "outsiders."[14] This diminished level of competition in turn enables members of the health care professions to raise their prices above what they would be in a free market.

The existence of medical monopolies has created a number of paradoxes in American life. Here is one of my favorite examples. If I want a pickled egg at 2:00 AM, I can get one. (My local supermarket, responding to competitive pressures, is open at that time; and in trying to satisfy regional tastes, it stocks pickled eggs.) If, on the other hand, it is noon on a weekday and I break my leg, I might spend several hours waiting (in a considerable amount of pain) in a hospital emergency room before a doctor will see me. If I have an ear infection that is low grade but irksome, my doctor (or rather, my doctor's nurse) might inform me that I will have to wait several days before I can have an appointment, and when I show up for this appointment, I might find that I am asked to wait an hour or two beyond the appointed time before my doctor is ready to see me.

Someone new to America might take all this as evidence that Americans value pickled eggs more than they value people's health. They would be wrong, though. It's not that Americans value eggs more than health, it's that we have granted monopoly status to health care professionals, while requiring those who provide pickled eggs (and other comestibles) to engage in open and ruthless competition with each other.

The health care crisis has, of course, causes other than the ones I have listed above. Nevertheless, I think I have listed some of the most significant ones, and I think that my list shows how diverse the factors are that affect the cost of health care.

Can National Health Insurance
Solve the Health Care Crisis?

Before discussing it any further, I should take a moment to clarify what I have in mind when I talk about national health insurance. After all, "national health insurance" means different things to different people. For some, it means nothing less than nationalization of all health care services, and for others, national health insurance would involve nothing more than the provision by the federal government of health insurance policies to those who cannot otherwise afford them.

The national health insurance plan I have in mind—and the one I shall be referring to in what follows—works along the lines of Medicare, except that it would apply to everyone and not just to people over 65. More precisely, almost everyone would be required to make a contribution to a national health care fund. The size of the contribution would presumably depend on one's income, and those with little or no income would receive coverage without being asked to make a contribution. When people needed medical attention, they would go to the private health care provider of their choice. The government would pay all or part (depending on how we fill in the details of the plan) of the treatment they received. People would not be restricted to using only this national health insurance; instead, people could, if they wished, purchase private health insurance policies, or could purchase "gap filling" policies that would supplement their national health insurance policies. (Of course, their doing so would not relieve them of their obligation to share the costs of national health insurance.) My guess—and it is only a guess—is that, given the current political landscape of America, this particular form of national health insurance has as good a chance as any (i.e., as any "real" form of national health insurance) of being enacted.

The question we must now address is this: Is there any reason to think that the form of national health insurance just described (or some variant of it) would alleviate the crisis in American health care?

There are those who will argue that in a national health insurance plan like the one described above, there are steps that the government could take to force down the cost of medical care. Let me describe three such steps.

First, some have suggested that the government, by carefully administering its insurance plan, could dramatically cut the cost of health care. Along these lines, one group, Physicians for a National Health Program, has proposed a national health insurance plan that it says would save about $50 billion in health care costs each year by slashing administrative costs.[15]

My primary reaction to this line of argument is this: It boggles my mind that anyone would think that by letting the government administer a program we could reduce administrative costs or make administration of the program more efficient. I say this because I normally don't think of the government as a prime example of efficient management; I normally don't think of the government as a vicious cost-cutter; and I normally don't think of the government as being ruthless in laying off employees that are "deadwood" in an organization. Instead, I see the government as the embodiment of inefficiency and waste, and I don't think I am alone in holding such views.

Indeed, to get a better idea of the government's chances of effectively administering a national health insurance plan, we should examine those insurance programs that it is currently either administering or closely associated with. Names like Social Security, Medicare, and the Federal Savings and Loan Insurance Corporation (FSLIC) come to mind. None of these is what a business school professor would hold up to his or her students as an example of efficient management. These are, after all, programs that have been saved from extinction only by massive and, in some cases, repeated government bailouts.

It is conceivable that national health insurance would not head in the same direction as, say, the FSLIC, but I have my doubts. For the fact of the matter is that government officials simply aren't very good businessmen. The reason for this is twofold: Government officials, typically, are sheltered from the consequences of their mistakes, and they typically aren't rewarded for innovation. The private businessman, on the other hand, is fully exposed to the consequences of the mistakes he or she makes, and if a businessman makes enough mistakes, he or she is not only out of a job, but possibly even bankrupt. Furthermore, a private businessman has a strong incentive to try harder and do better at whatever he or she is doing, for if this businessman can come up with what the customers judge to be a superior product or service, he or she will profit handsomely.

Furthermore, when the government operates a business, it typically cannot do so in a businesslike fashion. Politics will almost invariably play a role in decision making, and with politics comes a whole range of inefficiencies.

The above comments may sound like I have an inordinate amount of faith in private insurance companies, or that I take them to be without waste and incompetence. I do not; I regard them in many respects as flawed. My only claim is that most private insurance companies are managed far more efficiently than are nearly all government agencies.

A second step that the government might take to drive down health care costs (some will suggest) is to dictate prices to the health care community. If national health insurance became the dominant form of health insurance, the government could reduce health care costs by simply refusing to pay more than a certain amount for a certain operation, more than a certain amount for a certain drug, for more than a certain number of days in the hospital for people with a certain ailment, and so forth. Health care providers would have little choice, the argument goes, but to swallow the cost cuts.

Indeed, not only is the government role as price-dictator imaginable, but it is a role that the government has had some practice playing. It has, for example, tried this sort of thing with Medicare and Medicaid.

Could the government drive down the cost of medical services simply by reducing the amount it would pay for these services? I am certain it could. Nevertheless, I don't think the consequences of this cost reduction would be desirable.

Notice, after all, that in dictating prices in the manner described, the government is, in effect, imposing price controls on the medical community. Price controls can reduce costs, but they almost always have a number of undesirable side effects. Most strikingly, price controls can result in shortages of goods and services and in a decline in the quality of those goods and services that are available.

Suppose, for example, that the government tells the medical community that it will pay no more than $300 for a certain treatment. Suppose that most doctors in the medical community are unwilling to perform the treatment in question for under $500. (Perhaps it costs them $500 to perform the treatment, so that unless they are paid $500, they will lose money; or perhaps they judge that there are other treatments they can specialize in that will be more profitable to them.) In such a case, patients may find it difficult to get the treatment in question. (Along these same lines, when the government of Poland imposed strict price controls, Poland had the cheapest pork in Europe. The problem was that stores never had any to sell.)

Indeed, if the government sufficiently undercompensates health care providers, it might drive them out of business, drive them into other fields, or drive them to move to countries where their skills are more highly rewarded. Who will remain to provide health care? Most likely, those whose skills have a lower market value. The problem, of course, is that these won't necessarily be the people we want providing our health care. Thus, although dictating prices to the health care community might succeed in reducing the cost of health care, it can be argued that it will simultaneously reduce both the availability and quality of health care. It is far from clear that this is a tradeoff we will want to make.

This brings us to a third step by which some might claim that the government, by administering a national health insurance plan, might reduce the cost of health care: The government can act vigorously to curb fraud. The claim is that government officials will be better at detecting and prosecuting fraud than private insurers, and that because of this, national health insurance could provide the same coverage for significantly less money than private insurance plans.

My problem with this suggestion is that history indicates that government agencies generally aren't effective in preventing fraud—indeed, there is every reason to think that government agencies are a particularly ripe target for fraud. Again, a number of recent cases come to mind, including the scandal at the US Department of Housing and Urban Development (HUD), the looting of numerous savings and loans under the nose of government regulators, the ongoing parade of welfare fraud schemes, and so forth.

Of course, private businesses can also be victims of fraud. The thing to notice, though, is that the private businessman is generally more motivated to prevent and detect fraud than are government employees. The businessman, after all, is protecting his or her own money, whereas the civil servant is protecting someone else's money, viz., that of the taxpayers.

In summary, it seems that the government, acting as administrator of a national health insurance plan, would be limited in the steps it could take to decrease the cost of health care. Indeed, a case can be made that, rather than reducing health care costs, a national health insurance plan like the one described above could have the effect of raising them. It is true that the price increases might be indirect (e.g., they might take the form, not of higher medical bills, but rather of increased tax payments to fund periodic bailouts of the national health insurance plan), but they would nevertheless be real.

Before moving on, let me describe a different sort of argument that might be offered in support of national health insurance. In the argument in question, the claim is not that national health insurance would be a solution to the cost crisis in general, but rather that a limited form of it could deal with one aspect of the cost crisis, namely, the inability of large numbers of Americans to afford health insurance. The proposal, basically, is that the government should offer subsidized health insurance to those with low incomes. Under this plan, the poor might be able to get a health care policy that would, hypothetically, cost $3,000 from a private insurer for only $300, or might even be able to get the policy for free. The argument, in brief, is this: The poor are entitled to health care, and the best way to provide it is by having the government offer the poor subsidized health insurance policies.

The problem with the above argument is that both of its premises are questionable. In the first place, it is far from clear to me that the poor *are* entitled (in the strong sense of the word) to health care. This is a premise, of course, that most Americans would accept. It is important to realize, however, that not only are there Americans who reject it, but that for the first 150 years of our nation's existence, most Americans did reject it.

An attack on the above entitlement claim will not be attempted here. Instead, attention will be made to the second premise of the above argument, namely, that the best way the government can help the poor obtain health care is to provide them with subsidized health insurance policies.

Before we can determine whether government-issued policies would be the best way to help the poor obtain health care, we need to know what the alternatives are. Along these lines, I would like to suggest that we take a look at the following alternative: Instead of having the government "go into the insurance business" by writing health insurance policies for poor people, the government might give them vouchers that could be used to obtain health insurance policies from privately run insurance companies. The recipient of a voucher could "spend" it on the policy of his or her choice, presumably on the policy that best suited that person's needs.

Some will respond to this proposal by asking, "Why go to all the trouble of issuing vouchers? Why not just have the government write the policies itself?" The chief advantage of the voucher proposal, as I see it, is that it would put private insurance companies in competition with each other for the business of the poor, and would therefore improve the service and benefits they receive. If a company was rude to a voucher user, or lost the user's files, or didn't give the user much coverage for his or her money, the user in question could take his or her business elsewhere. The same is not true when we allow only one entity—namely, the government—to insure the poor. If the government is alone in providing a service and the user doesn't like the way the service is provided, too bad. Under this arrangement, the user cannot take his or her business elsewhere. (By way of analogy, the user may get poor service from his or her local shoe store, but just think of how bad the service would be if it were the only shoe store in town. And think of how much higher the prices would be.) In other words, a case can be made that under a voucher plan, the poor would get more coverage per tax dollar and be generally more satisfied than under a plan in which the government played the role of insurer.

The reader should not assume that the advocacy of the voucher plan, as described above, is unconditional. Instead, the voucher plan should be favored only if we decide that the government should take steps to guarantee health care for the poor.

Another Way to Deal with the Crisis

Above, an argument is made that there is little reason for thinking that the "obvious" form of national health insurance can alleviate the crisis in health care. Let me now address a new question: Is there something we can do to slow down or even halt soaring health care costs? I think there is, and in what follows, I will describe an alternative solution (a partial solution, to be more precise) to America's health care crisis. As will soon become evident, my solution takes us in the opposite direction, politically speaking, from national health insurance; I regard government involvement not as the key to solving the health care crisis, but as a major cause of the health care crisis.

The solution I have in mind is suggested by a reexamination of the causes of the crisis listed in Section 2. Of the four causes listed, the first two—namely, advances in medical technology and inelastic demand for health care—seem to be unavoidable.

The third cause—the "shield" effect—is avoidable. The government could, for example, mandate a rise in deductibles or copayments in private insurance policies, or it could enact a national health insurance plan that called on citizens to bear a much larger burden of health care costs. In doing so, the government would diminish the shield effect and would probably bring about some reduction in health care costs. The problem here is that the government would also force many Americans to forego various medical treatments. Without something to shield them from the costs of medical treatments, they would be unable to afford them. Most would argue that this "cure" for the health care crisis would be "worse than the disease."

This brings us to the fourth cause listed, the existence of medical monopolies. If I am right in thinking that these monopolies are a major factor in the high cost of health care, then it follows that by taking steps to weaken these monopolies—and thereby increase the amount of competition between health care providers—we can have a significant impact on America's health care crisis. This, then, is my alternative (partial) solution to soaring health care costs: take steps to undermine medical monopolies. (I am not, to be sure, the first person to suggest the desirability of weakening medical monopolies,[16] nor, for that matter, is this the first time I have made such a suggestion.[17])

One way in which we can combat rising health care costs is to take steps to increase the competition between the various health care professions. We accomplish this by "blurring" the lines that now demarcate the various professions. What we should do is allow people who are not medical doctors to perform various medical treatments that now fall within the domain of medical doctors. We should, for example, allow people who lack MDs to specialize in the treatment of sprained ankles. These "medical technicians"[18] would have a considerable knowledge of sprains, but they would know little about the rest of medicine. They would, for example, be ignorant of how to treat urinary tract infections, hair loss, or parrot fever. Other medical technicians might specialize in the treatment of ear infections; yet others might specialize in the treatment of hay fever or even in cosmetic surgery. These technicians would lack the general medical knowledge that medical doctors possess, but they would be competent to practice their little corner of medicine.

The reader should not make the mistake of thinking that I would severely restrict the treatments these technicians could perform. I would, for example, allow them to prescribe medicines or even to perform simple surgery—as long as the medicine or surgery in question were within their limited sphere of expertise.

If this proposal sounds too radical, the reader should remember that the American medical system currently allows a certain degree of "specialized medicine" to be practiced. For example, optometrists, podiatrists, and clinical psychologists treat eyes, feet, and minds, respectively, without benefit of an MD. They come close to what I have in mind when I speak of medical technicians. My proposal is simply that we encourage this sort of thing, that we allow non-MDs to treat more than eyes, feet, and minds.

What would we accomplish by allowing medical technicians to practice medicine? We would accomplish at least two things.

First, we would more efficiently employ our medical resources, for a case can be made that we are wasting our medical resources whenever we have a medical doctor—complete with many years of generalized medical training—spend his or her days looking at sprained ankles, treating cases of hay fever, removing warts, and doing a number of other things that in no way require the massive body of knowledge possessed by a medical doctor. It is, in other words, a waste to let a generalist do a job that a specialist could do with far less training and experience. Presumably, by using our medical resources more efficiently, we can cut health care costs. The services of someone with one year of limited training should be far less expensive than those of someone with a decade of generalized training.

To better appreciate the waste involved in the "let-an-MD-do-it" approach to medicine, consider this: If you want a ten-course gourmet meal, it makes all the sense in the world to hire a famous French chef to do the cooking. The job, after all, will require this chef's many years of experience and training. If, on the other hand, you only want a bunch of marshmallows roasted, it will be a waste for you to hire a famous French chef to do the roasting for you. After all, an eight-year-old could not only do the job in question, but would do it for far less money than the French chef would. Requiring that cases of hay fever be treated by people with massive amounts of medical training and experience is analogous to asking a famous French chef to roast a marshmallow. No wonder health care is so expensive in America.

A second thing we could accomplish by allowing a variety of medical technicians to practice medicine is that we would increase the amount of competition between health care providers. Since medical technicians would be relatively inexpensive to train, American medical schools could turn out vast numbers of them. When they began practicing medicine, they would compete not only with each other, but with medical doctors (or at any rate, with those medical doctors who chose to engage in treatments that did not call for their extensive medical knowledge). This competition would tend to push down the prices patients paid for health care, and it would thereby slow down increases in health care costs.

In a medical community that allowed the sort of specialization that I have described, the role of the medical doctor would probably change. Instead of using medical doctors to treat cases that involve minor ailments, we would "save them" for complex cases where their general medical knowledge is fully required.[19] In so doing, we could reduce the fees patients must pay to have minor ailments treated.

In my remarks above, I have paid attention primarily to medical doctors. I would be quite willing, however, to make similar proposals for other health professions. In dentistry, for example, I can imagine medical technicians who specialize in filling cavities (but who lack the general medical knowledge of dentists); in pharmacy, I can imagine use of medical technicians to perform many of the tasks for which pharmacists are over-qualified; and the same can be said of nursing and other health care professions.[20]

It is also important to realize that I do not claim that by undermining medical monopolies we can reduce America's overall medical bill. It may be that even with weakened medical monopolies, health care costs will continue to rise. (After all, as I said above, medical monopolies are only one of many causes of

America's health care crisis.) My claim is only that weakening medical monopolies is one important step we can take in trying to deal with soaring health care costs.

There are, to be sure, any number of criticisms that can be raised against my advocacy of the use of medical technicians. Let me describe some of them, together with my replies.

Criticism. In the plan described, people would need a variety of health care providers rather than a single doctor; after all, the person who was competent to treat your sprained ankle would not be competent to treat your ear infection, and so on. Aren't there advantages to having one person you can go to for almost any medical problem? Aren't there, in other words, advantages in the sort of one-stop medicine that our current system provides?

Reply. I agree that there can be dangers in overspecialization. I also agree that there are times when it will be not just desirable, but necessary, to go not to a medical technician, but to a full-fledged medical doctor. Suppose, for example, that I am plagued by a variety of mysterious symptoms. In such a case, I will want to consult a person possessing vast medical knowledge and experience, and I won't mind paying for his or her services.

Having admitted all this, let me go on to remind the reader that there are times when it would be nice to be able to resort to a medical technician if doing so would be more convenient and less expensive than going to a medical doctor. It would be nice, for example, if, when my son got what looked like an ear infection, I didn't have to wait (as sometimes has happened) two days before a medical doctor could see him. It would be nice if I could instead take him to my local "ear clinic," have a technician peek into his ear, and if necessary, prescribe the appropriate antibiotic; and it would be particularly nice if this technician's bill were substantially less than that of a medical doctor.

I should emphasize, by the way, that I am engaging in a bit of speculation when I talk about medical technicians who specialize in sprained ankles or ear infections. I don't really know

what specializations would arise if America became more toler-
ant of medical technicians. Presumably, the market for medical
services would decide such matters.

Criticism. Wouldn't the lower standard of training of medical technicians result in a reduction in the quality of health care?
In particular, wouldn't medical technicians be more likely to mis-
diagnose an ailment than would a medical doctor? What about
those cases in which what looks like an ear infection is really a
symptom of some rare and deadly tropical fever? Could a medical
technician be expected to spot the problem?

Reply. I admit that cases like the one just described can
occur. In short, I admit that the sort of specialization I am proposing will have certain costs. Before taking this as an admission
of defeat, though, the reader should realize that our current system of allowing only medical doctors to practice "everyday
medicine" has its costs as well. Because of their monopoly status,
as I have suggested above, medical doctors are both expensive
and difficult to see. The result is that many people postpone
treatment of ailments or let them go untreated. Such behavior, of
course, can cause a number of problems. We are left, then, with
an interesting empirical question: Which will provide the best
ratio of costs to benefits, the current system or the system I have
proposed? I am suggesting, of course, that the system I have
proposed, despite its flaws, would be a substantial improvement
on the current system.

Those who are worried about misdiagnoses by medical technicians should take a moment and ask whether they are likewise
worried about misdiagnoses by those "medical technicians" that
our system currently allows, e.g., optometrists, podiatrists, and
clinical psychologists. No doubt there are times when these practitioners make a mistaken diagnosis or improperly treat a patient,
but we should keep two things in mind. The first is that medical
doctors also make mistakes, and the second is that for many
Americans, the convenience and reduced cost of the above-named

"technicians" outweighs the slight risks associated with the "incomplete medical knowledge" these professionals possess.

Criticism. Under my proposal, wouldn't the incomes of medical doctors and other health care professionals fall? And wouldn't this in turn drive many health care professionals out of business and reduce the incentive for America's best and brightest students to choose careers in the health care professions?

Reply. I admit that my proposal would be bad news for many health care professionals; no one who enjoys the protection of a monopoly wants to face the prospect of competition. I also admit that if we were running the health care system for the benefit of health care professionals, my proposal would be extremely counterproductive. I would like to think, though, that we are running the health care system not for the benefit of health care professionals, but for the benefit of their patients.

Notice, by the way, that my proposal would not mean the end of medical doctors and other medical generalists, and those medical doctors who spent their days putting their extensive amount of knowledge and experience to use would find themselves well rewarded. Those medical doctors, on the other hand, who chose to use only a fraction of their knowledge (i.e., to waste their extensive training) would find themselves competing with a vast number of technicians, and they would suffer financially because of it.

What about the charge, though, that if the income of health care professionals declines, the quality of American medicine would likewise decline, since the best and brightest students would no longer be attracted to the health care professions? The problem with this line of thought is that it assumes that the best doctors are motivated primarily by money—that the best candidate for medical school is the one who is in it for the money. I question this assumption, and I think that I have company in doing so.

Americans, of course, generally oppose monopolies, and for good reason. They realize, in the first place, that the goods and services provided by monopolists are generally overpriced and of inferior quality, and that monopolists often abuse the power their monopolies give them. The puzzle is why Americans support the medical monopolies described above when there is every reason to think that these monopolies impose a number of costs on health care consumers.

In closing, let me remind the reader that in the above pages, I have not "refuted" national health insurance, nor have I given a complete defense of my "medical technician" proposal. What I hope I have accomplished is to encourage the reader not to give in to the common impulse to turn to the government to solve problems that "mere mortals" cannot solve. Indeed, when the reader detects in himself or herself such an impulse, the reader should recall (1) that the government is itself composed of "mere mortals" (including those merest of mortals, politicians and bureaucrats), and (2) that in many cases, the government, rather than being the solution to our problems, is in part to blame for them. To the extent that this is true, national health insurance, rather than being the solution to America's health care crisis, may only succeed in making the crisis that much worse.

Notes and References

[1]For a summary of the most important of these bills, *see* the Appendix of *National Health Insurance: Conflicting Goals and Policy Choices* (1980), Feder, J., Holahan, J., and Marmor, T., eds., The Urban Institute, Washington, DC.

[2]Esther, R. J. "Health-Care Costs for Businesses Soared in 1988...," *The Wall Street Journal* (henceforth identified as *WSJ*), 24 May 1989, A2.

[3]"Health Plan Costs Jumped...," *WSJ*, 31 January 1989, C20.

[4]Feinstein, S. "Outpatient Care...," *WSJ*, 30 May 1989, A1.

[5]"Firms' Benefit Costs in '89 Seen Rising Significantly," *WSJ*, 11 January 1989, A4.

[6]Karr, A. R. "National Health Plans Intrigue More Employers...," *WSJ*, 16 May 1989, A1. Along these lines, *see* also Ron Winslow's

"National Health Plan Wins Unlikely Backer: Business," *WSJ*, 5 April 1989, B1, and Albert R. Karr's "Please, Mr. Bush,...," *WSJ*, 31 January 1989, A1.

[7]Feinstein, S. "Workers Pay More...," *WSJ*, 11 July 1989, A1.

[8]Karr, A. R. and Carnevale, M. "Facing Off Over Health care Benefits," *WSJ*, 11 August 1989, B1.

[9]Davis, K. "National Health Insurance: A Proposal," *American Economic Review* **79**, 349-52.

[10]Waldholz, M. "Pediatricians Back National Health Insurance Plan," *WSJ*, 24 July 1989, B2.

[11]"Universal Health Insurance: Its Time Has Come," (1989) *The New England Journal of Medicine* **320**, 117-8.

[12]Brazda, J. F. (1989) "National Insurance Resurfaces," *Modern Health-Care*, 16 June 1989, 36.

[13]These generalizations all have exceptions, of course; what health care professionals are and aren't allowed to do depends on the state in which they practice, and changes from time to time.

[14]In saying this, I do not mean to suggest that doctors and other health care professionals face no competition whatsoever; they do, after all, compete with other doctors, and in some cases, with other health care professionals.

[15]Ruffenach, G. "Physician's Group Proposes National Health Program," *WSJ*, 12 January 1989, p. B1.

[16]*See*, for example, the section titled "Medical Licensure" in chapter IX of Milton Friedman's *Capitalism and Freedom* (The University of Chicago Press, Chicago, IL, 1962).

[17]*See*, for example, my "The Case for Physician-Dispensed Drugs," *Biomedical Ethics Reviews: 1989* (The Humana Press, Clifton, NJ, 1990).

[18]I would refer to these people as "medical specialists," but I think that this choice of terminology would be unwise. Most people, when they think of medical specialists, think of medical doctors who have gone on to acquire specialized knowledge in one area of medicine. My "medical technicians," on the other hand, acquire specialized knowledge of medicine without first acquiring general medical knowledge of the sort possessed by a medical doctor. The reader should also be careful not to confuse my medical technicians with those health care professionals known as medical technologists.

[19]Actually, I am being inaccurate in speaking of "saving them," for in the setup I have in mind, anyone who wanted to could visit a medical doctor. He or she would presumably have to pay for the privilege, though.

[20]In some states, of course, these suggestions have already been implemented. I am suggesting that they be implemented in other states, and that those states that are currently experimenting with "medical technicians" expand these experiments.

National Health Insurance

An Ethical Assessment

Arthur J. Dyck and James S. DeLaney

Introduction

For more than 75 years, some kind of national health insurance (NHI) has been proposed for the United States. Rashi Fein, who has documented the debates concerning these proposals, finds the staying power of NHI to be quite remarkable. One can, he believes, draw two opposite conclusions from this: "that, since the conception of NHI has survived for three-quarters of a century, it can't be all bad; or that since it has not been enacted over such a span of time, it must be fatally flawed."[1] But whatever the merits or flaws of any particular form of NHI, the problem that prompts the debate about the appropriate form of government persists: it is the problem of an increasing number of "have-nots;"[2] 37 million Americans were without any private or public health insurance in early 1987; prices for health insurance continue to rise at between one and one-half and two times the rate of the consumer price index.[3]

It is not possible in this brief essay to consider all the current responses to the fact that, for a variety of reasons, a considerable number of Americans increasingly experience price-rationing,

From: *Biomedical Ethics Reviews • 1990*
Eds.: J. Humber & R. Almeder ©1991 The Humana Press Inc., Clifton, NJ

sometimes to the point of being priced right out of the health care system in time of need.[4] Responses by philosophers tend to take the form of articulating a concept of justice that serves either to justify, or defend against comprehensive government programs to provide comprehensive health care.[5]

Such responses, though, are incomplete at best. As Fein has amply documented, groups of legislators who agree that the national government should help guarantee universal health care coverage have, nevertheless, defeated one another's proposals; one can agree that a particular goal is just and yet disagree on how to achieve that goal, or as to whether the goal can even be achieved. Just goals will remain unrealized so long as there is sufficiently persuasive opposition to any of the specific means proposed to reach these goals. To the best of our knowledge, no philosopher addressing the subect of health care is morally indifferent about anyone who may suffer or die for lack of care; the case for comprehensive, government-sponsored health programs does not follow from a genuine concern for those who fail to receive care necessary for their health and very survival. The debate over NHI is not only a debate over what justice demands, but also over how these demands can and should best be satisfied.

To assess, then, the cogency and relative strength of arguments for and against NHI, we have deemed it most helpful, perhaps even necessary, to examine concrete proposals. We have chosen from among those proposals submitted for public debate and with the intention of seeing them through to legislative enactment and implementation. These are proposals that address actual and anticipated obections various sectors have, or may have, to certain kinds of involvement by the national government in the provision of health care. This sensitivity to American attitudes precludes advocacy of a national health service, as in Britain, which, unlike national health insurance, makes health providers employees of the national government, working in its facilities. The proposals we are examining also take into account that the US already has a 400 billion dollar health system in place. This system includes government involvement, such as financial support for medical research and medical education, programs such as Medicare and Medicaid, and tax incentives for employers who provide employee health benefits.

As Fein observes, the existing American ethos and government involvement serve as constraints that "rule out widely different options: a national health service and an unbridled free market."[6] That, however, does not mean that specific proposals do not draw fire from concepts of justice favoring one or the other of these options. It does mean, however, that such concepts of justice have rather indirect and unpredictable influence on what legislators have so far enacted and debated in the US. Again, only a consideration of concrete proposals will bring to the surface the competing concepts of justice that underlie these policies and debates.

The purpose of our essay, then, is to offer an ethical assessment of the most cogent arguments for and against a national policy designed to provide universal health care insurance within the US. This we do by analyzing some specific policy proposals, selected for reasons already briefly described above. By policy, we mean a particular goal, or set of goals, with specified means to that goal, or set of goals. But what do we mean by an ethical assessment?

In studying proposals for NHI, it became apparent that the reasons given for having NHI rest on a moral basis. Appeals are made to "need," "equity," "fairness," and "egalitarianism." Facts are cited: to indicate what is demanded in the name of needs, or equity, or fairness, or egalitarianism; to indicate by whom these demands are to be met; and to whom these demands are owed. These proposals are not content with things as they are—things are not right as they are. Why not? Because some things owed to certain individuals and groups are not being provided. Someone, in these instances the whole society by way of its government, is strictly obliged to provide what is not being provided. The structure of such an argument is precisely the one expressed in the very basic and historically early definition of justice, namely, "giving to each their due."[7] And what is owed or due someone is now more typically specified in the language of rights. A clear example is found in the writings of John Stuart Mill: "Justice implies something that is not only right to do and wrong not to do, but which some individual person can claim from us as his moral right."[8] Furthermore, these claims are made on behalf of what Mill calls "the essentials of human well-being."[9]

To begin with then, this is a study in ethics because it renders explicit what is mostly implicit in the debate over NHI, namely the competing views of what justice demands. And, as indicated above, in this study, we will be interpreting this debate about what justice demands as a debate about rights.

But policy proposals invite ethical evaluation in other ways. If a policy proposal is to be persuasive, it will be compelled to overcome at least three kinds of objections to its acceptance and implementation.

1. That the existing policy, or another policy other than the one being proposed, is more just, or at least no less just than what is being proposed;
2. that, whatever the merit of the policy being proposed, it cannot succeed, or the chance that it will is very remote; and
3. that the proposal is, for whatever reason, cognitively flawed, or provides no predictable or reliable process for avoiding serious cognitive errors.

In summary, our ethical assessment will evaluate four concrete proposals of NHI from the perspective of:

1. what justice demands;
2. whether each is in any sense necessary, given the alternatives, actual or possible;
3. how likely each is to succeed; and
4. how cognitively sound each is.

These are not arbitrary criteria for evaluation; they all occur within the debate over NHI. Each will be classified and defined in use rather than now, in advance.

"A National Health Program for the United States: A Physician's Proposal"[10]

This proposal is put forward by Physicians for a National Health Program. It is written by a committee co-chaired by David Himmelstein and Steffie Woolhandler. The authors have provided a concise summary as follows:

> We propose a national health program that would (1) fully cover everyone under a single comprehensive insurance program; (2) pay hospitals and nursing homes a total (global) amount to cover all operating expenses; (3) fund capital costs through separate appropriations; (4) pay for physician services and ambulatory services in any of three ways: through fee-for-service payments with a simple fee schedule and mandatory acceptance of the national health program payment for a service or procedure (assignment), through global budgets for hospitals and clinics employing salaried physicians, or on a per capita basis (capitation); (5) be funded, at least initially, from the same sources as at present, but with all payments disbursed from a single pool; and (6) contain costs through savings in billing and bureaucracy, improved health planning, and the ability of the national health program, as the single payer for services, to establish overall spending limits.[11]

What Justice Demands

These authors are calling on the national government to adopt their program as an alternative to the present health care system. Why? Because they view the present system as failing. They cite as its failures:

1. many people in need who are denied access to health care;
2. expense, inefficiencies, and a growing bureaucracy;
3. pressures that threaten the traditions of medical practice in the form of cost control measures, competition, and the quest for profit; and
4. fear of financial ruin in patients already suffering.

That health care should be allocated in accordance with need is one widely held conception of justice as applied to health care.[12] To argue that governments should mandate a program that assures such an allocation is to treat such needs as the basis for a right of access to the health care appropriate to meet those needs as they arise. This, in turn, implies that a society, through its governing bodies, has a strict obligation to provide the resources necessary to make such access possible. The quality and amount of care that should be universally available will be taken up below in discussing Himmelstein's and Woolhandler's views on how such decisions are to be made.

Their concern for those who suffer illness does not end strictly with medical needs. The cost of care, particularly for those uninsured, but also for those underinsured, can be a financial disaster. Prospectively, this can frighten some people from seeking access or continuing in care; retrospectively, this can leave someone utterly bankrupt, completely poverty-stricken. The authors are not explicit about it, but they seem to be invoking society's more general social mandate to enlist government aid for all those who are too poor to meet their basic needs for food, shelter, and the like. This is a more universal application of justice as a strict obligation to meet certain human needs. Overcoming illness and receiving care for illness are not to be left to the existing philanthropic impulses and charitable organizations of the private sector.

The costs and inefficiencies of the present health care system, as well as the lack of insurance or sufficient money for millions of patients, are unjust for health care providers. The standards of care are compromised by providing less care for the uninsured, avoiding some procedures, consultations, and costly medications. The use of diagnosis related groups (DRGs) to cut costs places physicians between the demands of administrators for early discharge and the needs of elderly patients who cannot receive care at home. Himmelstein and Woolhandler mention also the concern for the "bottom line" in HMOs and the general lack of ensuring basic services in public health, such as prenatal care and immunizations for everyone. Again, they are mounting a moral argument that the medical profession cannot give the care it ought to give. The resulting unmet needs of patients and the inequalities in the treatment when it is offered, i.e., substandard care for the uninsured and the underinsured elderly, are conditions that society is strictly obliged to rectify. It is a matter of justice extending also to the rights of health care providers to be able to practice medicine with a clear conscience. Himmelstein and Woolhandler do speak of their plan as the road to a rational and humane health care system, and as a way to transform disaffection with what exists into a vision of what might be.[13]

So far we have noted that the costliness of medical care is cited as playing a role in unjustly denying care to some, either by denying access or by providing less than what is best or needed. In effect, then, the present system practices what has been called "price-rationing."[14] However, Himmelstein and Woolhandler are concerned with the escalating costs of medicine relative to the total goods and services available to the US economy and US government. They suggest the following limit to spending: "The total expenditure would be set at the same proportion of the GNP as health costs represented in the year preceding."[15] Since they do not give a rationale for setting a limit or setting this particular limit, the assumption would seem to be that the proportion of the nation's wealth now being spent is about right. Existing injustices will be rectified without an additional proportion of GNP provided that the national government is the single payer of costs and requires all health services and health care providers to stay within the budgets allocated. The expectation is that costs will be reduced by this change in administration and in enacting incentives to reduce "excessive care."[16] What Himmelstein and Woolhandler hope to gain is clear: a reduction of what they regard as injustices in the health care system with no more increases in the wealth allocated to that system. Unclear are the reasons that the present proportions of health expenditures are as they should be. Nevertheless, this cap on expenditures suggested by the authors is put forward as an obligation of the national government and, as such, rests on an implicit conception of distributive justice, i.e., of how the nation's resources ought to be allocated. Of course, since the authors expect the national government to pay its bills with tax dollars, there is an obligation not to waste such money and to look for savings. That, however, does not by itself yield clues as to how much money ought to be spent on health care.

Necessity

Himmelstein and Woolhandler do not consider other propoals to change the US health care system. Their plan is presented as the needed alternative to the present system. Their plan will rectify the injustices that specify the ways in which the present health care system is failing. The proposed alternative is not compared to other alternatives, except to note favorably its resemblance to Canada's national insurance policy. They do note without elaboration that "patchwork reforms succeed only in exchanging old problems for new ones."[17] Nothing short of a comprehensive national health program will do. The coverage this plan envisages is one in which:

> Everyone is included in a single public plan covering all medically necessary services, including acute, rehabilitative, long-term, and home care; mental health services; dental services; occupational health care; prescription drugs and medical supplies; and preventive and public health measures.[18]

Chance of Success

Himmelstein and Woolhandler are aware that their proposal will be judged as to its potential to succeed. And, furthermore, this potential success will be judged with respect to whether the goals of the program can be achieved and whether it will gain sufficient support, socially and politically, to be adopted and tested for its efficacy.

Himmelstein and Woolhandler believe that they have provided for meeting the medical needs of those individuals presently without access to care because they have a federally mandated and funded program for universal coverage, and this coverage is comprehensive, as noted above, and does not erect the present financial barriers to vital care through patient copayments and deductibles. These barriers are eliminated.

Himmelstein and Woolhandler cite the Canadian system of health care as evidence that comprehensive coverage and the removal of copayments and deductibles do not drive up the cost of health care. However, for this to be true it is necessary to have global prospective budgets, to sort out necessary from unnecessary care, and to cut administrative costs by insisting on the government as the single payer. Again, Canada is taken as an example of keeping down costs through the use of global prospective budgets and by having the government be the single payer.[19] And to top it off, they propose a cap on federal spending for their global budgets, which, if adopted, would of necessity keep costs down.

But what gives them any reason to expect that their goals, and the suggested means to attain them, will become policy for the US government? One reason they give us is that almost every poll of public opinion in the US over the past thirty years has shown "that the great majority of Americans support a universal, comprehensive, publicly administered national health program."[20] Not only the public approves a national comprehensive program, but physicians (56%) as well.[21] With respect to physician approval also, Himmelstein and Woolhandler invite acceptance from them by proposing flexibility on the method of paying for services, whether by fees, salaries, or capitation, leaving all three options in place to be chosen as appropriate to the health care settings. In favoring state and local administration of health services, fears of a distanced bureaucracy are considerably laid to rest. They, however, do not make this point explicitly. Also, patients would have a free choice of providers. This would reduce opposition as well.

Cognitive Processes

Early on in their proposal, the physician authors, Himmelstein and Woolhandler identify themselves as such: "We are physicians active in the full range of medical endeavors."[22] In this and the following two paragraphs, they continue to describe the range and nature of their experiences in the US health care system and the problems they see and the solutions they envisage as a result. They anchor their whole proposal by appealing to their credentials and their expertise. Of course, as we have observed heretofore, they do not rest their whole case on their authority and their own expertise.

In addition to calling for accountability to the electorate at both the federal and state level, they suggest local participation in planning and evaluating medical services: "Boards of experts and the community representatives would determine which services were unnecessary or ineffective, and these would be excluded from coverage."[23] Given this provision, the authors are certainly not claiming that medical expertise alone, or as such, should be granted the authority to specify which medical services are necessary and which are unnecessary to cover, using federal money. Of course, the money being used is money from taxpayers. Nevertheless, there is comitment to a public dialogue and to forums in which that dialogue take place directly, as well as among elected officials accountable to the public. At the same time, it should not be overlooked that the authors advocate a cap on federal spending. This sets a definite limit to public deliberations and their latitude in influencing those global budgets and their allocations to states and local communities. The authority for that provision is not rendered explicit.

"Health Insurance for the Nation's Poor"[24]

We turn now to a much more modest proposal for national health insurance offered by Uwe E. Reinhardt, a professor of political economy at Princeton University. We do so now because it is a plan presented on January 29, 1987 before the Senate Finance Subcommittee on Health, by someone who also favors a health insurance scheme similar to Canada's, and yet, who argues for the necessity of a much more limited change in the US health care system. Reinhardt's reasons for advocating what Himmelstein and Woolhandler might wish to call "patchwork reform" will serve to call into question a number of the arguments they use to defend their plan (we will refer to that plan as Plan 1 and Reinhardt's as Plan 2). Proceeding in this way will document the difficulties NHI proposals face in the continuing public debate in the US.

Reinhardt's plan has the following essential features:

1. federal fail-safe insurance for any American who does not have adequate private health insurance coverage;
2. nonemergency care fron a limited number of local health maintenance organizations (HMOs);
3. HMOs bidding competitively for the right to serve federally insured patients;
4. the quality of care externally monitored and patients able to vote with their feet;
5. emergency care available from the nearest provider and that care adequately compensated; and
6. financing of the federal program based on ability to pay.

The proposed fail-safe program would absorb Medicaid.[25]

What Justice Demands

Like the authors of Plan 1, Reinhardt begins to build his case for his federal program by pointing to the increasing number of uninsured individuals in the US. But the problem of unmet medical needs is not simply the numbers involved. Until the early 1980s, the uninsured could obtain relief for truly serious conditions from neighborhood hospitals. The cost of "uncompensated care was recovered from the insurers of privately insured patients. In this situation, not found elsewhere in the industrialized world, Reinhardt observes that:

> ...our system thrust indigent patients into the status of health care beggars forced in case of illness to cast about for some provider's *noblesse oblige*. Furthermore, it thrust on hospitals the unwanted task of collecting hidden taxes through a pin-the-tail-on-the-donkey game by which the cost of indigent care was stuck willy-nilly on the tail of paying patients (or their insurers and employers). Even so, it can probably be said that few Americans died or suffered severely for want of critically needed medical care, if they really tried hard enough to procure it.[26]

But this situation is not now sustainable. The present predilection for prospective compensation, competitively set, not only destroys the hidden tax base, but also puts the health of the poor at risk. This new situation poses the question as to whether people are willing to be taxed to assure that the health needs of the poor are met. It poses as well the question as to "whether, at long last, all of America's poor ought to be granted some basic entitlement to health care."[27]

Reinhardt makes it very clear that he is talking about demands of justice. He notes that "Americans love to luxuriate in the thought that theirs is a society with strong egalitarian traditions."[28] Whereas that holds for economic opportunities that draw people to the US from all over the world, he believes that no one can imagine that the situation now pertaining, nor the one pertaining in the early eighties, can be called "an egalitarian distribution of health services."[29] To nail down his contention, he reports two examples that occurred during 1985. One is of an accident victim lying unconscious in a Florida hospital that has no neurosurgeon while two larger hospitals with neurosurgeons refuse to accept him because there is no assurance that the bill will be paid. The other is of a comatose three-year-old girl who is refused by two hospitals in South Carolina because her family has no health insurance; a hospital 100 miles away finally admits her. Concluding his direct appeal to the American conscience, Reinhardt declares that:

> Such vignettes would be quite unconceivable in neighboring Canada or in Continental Europe, where 90% or more of all citizens share a universal health system accessible to all on equal terms, and only a well-to-do minority of 5% or so chooses to opt out...The vignettes take on added poignancy when it is recalled that the American health sector is actually plagued by a surplus of both doctors and hospital beds, and that this country spends a higher proportion of its GNP on health care than does any other country...In fact, few if any countries in the industrialized world can rival the United States in the best parts of its health care system, and few, if any, match it or would want to match it in its worst.[30]

Reinhardt shares the concern of the authors of Plan 1 for preserving the freedom of physicians to practice medicine in accord with their "professional code of ethics," free from the threats to their economic security and the detailed monitoring by external lay boards, which would be their lot if medical care were left totally to the free market. And, as Reinhardt points out, Canadian physicians and hospitals are "less directly controlled, clinically more unfettered; and financially more secure" than would be the case for their American counterparts if a free market system were to prevail.[31]

Necessity

The reasons for changing from the present US health care system and adopting Reinhardt's proposal focus, as we have seen, on the unmet needs and health risks of the poor and those otherwise uninsured or underinsured. Reinhardt's analysis also documents why these unmet needs and health risks will keep escalating unless, and until, his proposed remedy is adopted.

But, unlike Himmelstein and Woolhandler, Reinhardt takes account of alternatives to his plan. Indeed, he favors an NHI scheme of the sort suggested in Plan 1. Why then does he settle for much less? Because of what he calls a "viable policy option." Such an option is "one which stands a reasonable chance of being legislated and implemented."[32] Plan 1 is, for Reinhardt, not a viable option. He argues that "devising a viable policy option for the nation's poor" would be off to "a solid start if it were to accept as a more or less permanent policy parameter that this country is unlikely ever to implement in practice the lofty egalitarian precepts it professes during public debate on health policy."[33] Reinhardt believes that efforts to help the poor have been stymied by protestations that the poor have the best of health care. The best is so expensive that often nothing has been done. Thus, Reinhardt does not expect Plan 1 to succeed; he has hope for Plan 2. He regards his plan as an "improvement" over what obtains now, since it "would be an honest two-tiered health care system that would grant the poor dignified access to a humanely endowed bottom tier."[34]

Reinhardt also regards two other possible policies as not viable: policies that mandate employer-provided insurance and policies that rely on a free market. Since Reinhardt's major arguments against advocacy of Plan 1 and these latter two types of policies focus on the utter improbability of their enactment and implementation, we turn now to his explicit reasoning with respect to what will and will not succeed as a health policy for the United States.

Chance of Success

Although Reinhardt's major focus is on the likelihood that a given plan to provide health care for the poor will pass muster politically and be adopted, he also attends to the likelihood that a given plan will attain the goals it sets for itself.

Reinhardt is convinced that NHI, such as proposed by Plan 1, will not become national policy, certainly not in the foreseeable future, perhaps never. His main argument is based on his assessment of "realistic policy parameters."[35] To be viable, a policy must stay within such policy parameters. He defines a policy parameter as "a cultural, political, economic, or administrative constraint so immovable within the time frame of the proposed policy as to approximate a state of nature."[36] Reinhardt then takes up what he regards as the most important policy parameters constraining public policy on indigent care: (1) the national ethos; (2) the legislative process; and (3) fiscal constraints.[37]

The National Ethos

Americans think of themselves as egalitarian. This does apply to economic opportunity, as far as Reinhardt is concerned, but offering greater economic opportunity than other countries does not imply an egalitarian distribution of basic human services. Rather, Reinhardt maintains, a belief in equal opportunities to advance economically is likely to lead one "to favor an inegalitarian distribution of economic privilege, because ignorance and poverty are then viewed as products primarily of sloth, rather than of bad luck."[38]

Reinhardt does not view the inegalitarian distribution of health services as maliciously intended. If these service were less expensive, they would undoubtedly be shared by the well-to-do. But it is very difficult to move the well-to-do toward sacrificing some of the freedom of choice and superior amenities now enjoyed, given the societal view that success and poverty primarily result from free choice rather than mere good or bad fortune, respectively. There is an argument to support this unwillingness to guarantee strictly egalitarian access to all of the health care available. Reinhardt states it succinctly:

> The regulatory structures that must be erected to enforce socialization of a commodity inevitably blunt the innovative edge of the system that produces the commodity. Thus... enforcement of greater equality in the distribution of human services would come at the expense of technical progress in the production and delivery of these services.[39]

Based on international comparisons, Reinhardt finds that argument "hard to quarrel with."[40] Plan 1 is vulnerable to this argument on at least two counts: (1) it would keep federal spending constant relative to the GNP; and (2) it would create boards to specify what constitutes "unnecessary care." In the first instance, innovation may be reduced by lack of funding. In the second instance, innovation may be stymied because its "necessity" is uncertain, or not perceived. Of course, the problem of money is not so acute whenever the GNP experiences substantial increases. But the regulatory problem remains and the authors of Plan 1, Himmelstein and Woolhandler have not explicitly offered reasons why anyone can expect their plan, once implemented, is immune to it. That means that those who benefit from American medical care at present have a reason to fear some of the changes suggested in Plan 1 because the overall quality of medical care and its delivery are very likely subect to decline.

Himmelstein and Woolhandler do have a reply to this last appeal to fear of change among Americans. As noted above, they cite public opinion polls as evidence that the majority of Americans support a program like Plan 1, including the Canadian program. But Reinhardt is not moved by that survey data.

The Legislative Process

Even if there were to be a survey that would convince legislative representatives that the American public favors an egalitarian distribution of health care, Reinhardt contends that legislative action to bring about such a distribution would still not be possible. Reinhardt singles out the moneyed interests of three groups: trade unions, trade associations of health care providers, and health insurers. Although these groups do not have the power to dictate public health policy, each of them has been able to veto legislative proposals it opposes. Reinhardt draws from this the following very strong implications regarding what it takes to render viable a health program for the nation's poor: it "must not detract perceptibly from the bottom lines of health care providers and insurers, and...its chance for implementation is enhanced if it adds to these bottoms lines."[41] Furthermore, Reinhardt says of this policy parameter, that it is "virtually immutable even in the longer run."[42]

If Reinhardt is correct even about the power of insurers, Plan 1 can in no way be considered a viable policy option. After all, the suggestion in Plan 1 is that all private health insurance be phased out entirely. And, so far, no plan of this kind has passed muster. However, that by itself, is not a decisive argument with respect to what the future will bring. Canada did abolish private insurance indicating that it can be done, but the US government will have to reckon with larger, more powerful insurers and large groups that may not be ready to exchange what they have for what may yield them less.

Fiscal Constraints

Reinhardt takes note of the federal government's quest to reduce its deficits and to do so without raising taxes. Because of this, he does not expect a significant contribution from the federal government toward solving the problems of the uninsured any time within the next several years, other than to ignore hidden taxes on business. An example of that would be to mandate employer-paid health insurance. Reinhardt was certainly right to anticipate proposals of that kind. In 1989, several have been submitted for scrutiny, including one in the Senate and House.[43] We will be examining one such below.[44]

Although marketing private business firms to provide health insurance for all employees may be one of the only viable ways to help the uninsured in the near future, Reinhardt is critical of any policy that relies on hidden taxation. Hidden taxes are "invariably unfair and inefficient."[45] For example, businesses have to pass on the cost of mandated health insurance to their customers; failing that, they would have to try to reduce wages. Reducing wages, if successful, would constitute a regressive tax on employees. Those in the lower wage brackets could be seriously impoverished by this.

Reinhardt's major reasons for rejecting a free market approach to health care were pinpointed earlier in connection with what justice demands. But Reinhardt believes that physicians and hospitals would fight against market forces if they were fully unleashed. They would do so because the free market "is actually a tortuous instrument designed to visit on them daily uncertainty, step-by-step monitoring of their clinical decisions by external lay boards, and morally vexing tradeoffs between economic security and their professional code of ethics."[46] Reinhardt wryly suggests that from such a vantage point the Canadian system would look good to American providers, since that setting is "less directly controlled, clinically more unfettered, and financially more secure" for physicians and hospitals.[47] Interestingly enough, the physician advocates of Plan 1, as we have observed above, already profess to experience what Reinhardt attributes to the free market, and already expect relief from a program much like the Canadian one. They would say to Reinhardt that tomorrow is already here and they have the backing of the majority of American physicians. However, these physicians will still need to reckon with the other constraints against moving to NHI in America and address them explicitly. Reinhardt, after all, has crafted his plan to work within these constraints; he is challenging any plan like Plan 1 or the free market to show how they expect to overcome those constraints, at least in the foreseeable future.

Reinhardt believes that a major advantage of his proposal within the American context is that it does not call for a full-fledged national health insurance program. Through appropriate calibration of the tax rate on adjusted gross income, most Americans could be motivated to seek private insurance. Also, his plan sufficiently compensates health care providers for care of the poor. His plan respects the "bottom lines" of providers and private insurers, and significant policy parameters. Thus, no one has egalitarianism forced on them.

What about the cost of Reinhardt's program given his own warnings regarding financial constraints? Taxes would be based on ability to pay and the taxes would be low for those with insurance. Reinhardt suggests that additional money could be raised by levies on the consumption of alcohol and tobacco and by a health care tax on gasoline, a major toxic pollutant. And one might also remove a major remaining tax shelter and no longer exclude employer-paid health insurance premiums from taxable income. This last suggestion invites opposition from trade unions and private insurers. However, Reinhardt wonders how long this nation can justify subsidizing health insurance even for upper-income business executives while appealing to budgetary constraints as a justification for excluding millions of poor families from basic health insurance coverage. In this case, Reinhardt sees the current tax subsidy to private health insurance as "neither equitable nor economically efficient."[48]

Cognitive Processes

It is not surprising to find Reinhardt relying, at times, on the authority of economists. Indeed, the whole plan is the product of an expert policy analyst, taking into account the economic and political factors that shape policy and largely determine what policies are adopted and implemented. Reinhardt is sensitive to the situations of health care providers and of those in need of health care. He is not, however, explicitly concerned to increase their ability and power to help define their needs, rights, and responsibilities, and to participate in how the health care system could best respond to these. As we have noted previously, the authors of Plan 1 also did not move much in this direction. Plans 1 and 2, therefore, do not meet the expectations expressed in the principles by which the American Public Health Associations (APHA) judges any proposal designed to provide "National Health Care."

Several of these principles are concerned with the improvement of cognitive processes: consumer and provider participation in ongoing planning and evaluation for the sake of improving the delivery of health services; education and training of health workers; and consumer education as to their rights and responsibilities.[49] We shall have more to say about cognitive processes at the conclusion of our essay.

"A Consumer-Choice Health Plan for the 1990s: Universal Health Insurance in a System Designed to Promote Quality and Economy"[50]

Suppose someone were to agree with Reinhardt that there are constraints within the US such that a national insurance proposal like Plan 1, and the Canadian plan so similar to it, are not viable and will not be enacted, certainly not in the foreseeable future. Suppose also that this same individual, or collection of individuals, were to agree with Reinhardt that everyone now uninsured or otherwise barred financially from health care should be insured in order to provide them access to health care. Would such agreement lead to an endorsement of Plan 2, or of a plan essentially indistinguishable from it? No, it would not. The evidence for such an assertion is not difficult to discover. The "Consumer-Choice Plan for the 1990s" (Plan 3) noted above and put forward by Alain Enthoven and Richard Kronick differs from Plan 2, but the reasons for proposing it include those shared by Reinhardt, as indicated above. And so we wish to discuss Plan 3 because it represents an attempt to formulate a politically viable and socially acceptable system of universal health care coverage, and yet in ways that still invite debate. Thus, our understanding of the continuing debate over NHI, and the failure to attain it so far in the US, should be increased.

Briefly sumarized, Plan 3 has the following major elements in it:

1. "Everyone not now covered by an existing public program would be enabled to buy affordable subsidized coverage, either through their employers, in the case of full-time employees, or through 'public sponsors,' in the case of the self-employed and all others;"[51]
2. employers are required to provide private insurance for all full-time employees, and their dependents not otherwise covered;
3. the State provides subsidized private insurance for all others, including employees of small businesses (less than 25 employees);
4. all plans include at least the basic benefits packages of the HMO Act, subject to tighter definitions and restrictions as deemed necessary to control costs;
5. there is no premium below the poverty level and a sliding scale up to 150% of the poverty level;
6. consumers otherwise pay 20% of the average premiums, and employers or the public sponsor, as the case may be, 80%, and no more than 8% of the payroll for small businesses; and
7. all deductibles and coinsurance are limited to 100% of the annual premium.

What Justice Demands

Like the authors of both Plans 1 and 2, Enthoven and Kronick are concerned about the number of Americans who have "no financial protection from medical expense." And, as they go on to say, "the present financing system is inflationary, unfair, and wasteful."[52] These negative features are also increasing, not abating. The inflationary trend contributes to deficits in the public sector, threatening the solvency of some industrial companies, and creating heavy burdens for many people. Unfairness in the medical care system is not limited to the fact that a number of individuals and groups are priced out of the system. Enthoven and Kronick note that close to two-thirds of all uninsured people are above the poverty level and more than two-thirds of uninsured adults are employed. When such persons are seriously ill, the cost of their care is largely borne by taxpayers, insured persons, or both. From this, Enthoven and Kronick conclude that "voluntarily or involuntarily, some people are taking a free ride. Those who can do so ought to contribute their fair share to their coverage and be insured."[54] What is more, they, as did Reinhardt, point to recent cost cutting measures by governments and employers that are drying up hospital resources for providing uncompensated care, and have led some hospitals to close their emergency rooms.

Waste in giving health care is, in a number of ways, a byproduct of the financial barriers to care. The uninsured, for example, obtain much of their primary care in the outpatient departments and emergency rooms of public hospitals rather than in the less expensive physicians' offices. Postponing care, and lack of preventive care, such as prenatal care, lead to more costly, serious illnesses, and to illnesses that could have been avoided. Additional inefficiencies that drive up costs contribute to situations in which the personal savings of the uninsured are depleted.[55]

The demands of justice, then, are the basis for what Enthoven and Kronick call the "two main goals" of their proposal: (1) "financial protection from health care expenses for all;" and (2) promotion of cost consciousness.[56]

Necessity

If you ask Enthoven and Kronick why it is necessary to change the present health care system, they would cite the demands of justice discussed above:

We cherish efficiency and fairness, but we have a system that is neither efficient nor fair. Very few Americans believe that other Americans should be deprived of needed care or subjected to extreme financial hardship because of an inability to pay.[57]

They go on to indicate that the American society lacks institutions to correct such inefficiency and unfairness. They regard this proposal as the remedy for what is lacking.

But what about other proposals now being put forward that purport to overcome the inefficiencies and unfairnesses with which Enthoven and Kronick are concerned? Why must their plan be adopted? Why not one of the other contenders, such as Plans 1 and 2 discussed above, or still others? In a word, theirs is the necessary plan because none of the others will work. Their plan has a reasonable chance to be adopted and to meet its goals; all others they consider, fail in one or both of these ways.

Chance of Success

One of the major reasons Enthoven and Kronick have for believing that their plan is more likely to be adopted than any of the others is that it is a "proposal for incremental change, a change compatible with American cultural preferences."[58] Like Reinhardt, they do not think Americans will sacrifice what they already have in the quality of care for socialized medicine, as in Britain. Nor do they expect existing health care systems to go out of business quietly and without a struggle. Like Reinhardt, they regard HMOs as partially successful in increasing efficiency and maintaining quality. Their plan, like Reinhardt's, creates incentives for increasing reliance on HMOs. But it does so differently.

Enthoven and Kronick do not explicitly discuss Reinhardt's proposal. However, one could expect them to take issue with Reinhardt's suggestion of a broad-based tax to pay for extending and absorbing Medicaid. By doing this, Reinhardt takes a step, however large or small in the eye of beholders contemplating its implementation, toward a more federally controlled health care system. And that, Enthoven and Kronick contend, is something Americans oppose, favoring as they do, "limited government and decentralization."[59] One suspects that the authors also favor these values, and do not see them only as social realities that help determine whether a given reform of the health care system will succeed. Be that as it may, our authors contend that these values, and the reluctance to permit governments to redistribute income in any radical way, predispose Americans toward their proposal, since it relies on the reform of incentives in the private sector and encourages those kinds of health care systems already achieving some level of efficiency.

Enthoven and Kronick believe that the state as a public sponsor of health insurance can achieve economies of scale, so that smaller, and even middle-sized employers would be able to purchase insurance at lower rates, with states acting as brokers. The states can also achieve greater administrative economies by using the agencies that currently obtain coverage for public employees. Moreover, through the use of public sponsors, everyone can be covered at more reasonable rates. Senator Kennedy's current proposal, mandating employer-provided health insurance, lacks the provision of public sponsors, and will not, Enthoven and Kronick argue, cover all of those presently uninsured.

Enthoven and Kronick's plan raises money for these public sponsors by having every individual or family above the poverty line, and every employer, contribute some money to the federal government to be used to support the state administered public sponsors. Employees would only be tax exempt for their 80% contribution and this too would help finance the public sponsors. This provision contrasts with a Massachusetts proposal for mandating employer-provided health insurance that cannot use federal tax laws to generate the funds to assure health care for everyone.[60]

Enthoven and Kronick realize that their proposal would mean that the minimum wage would, in effect, be raised by 8%. They suggest ways in which this effect could be mitigated.[61] No plan to extend health coverage can be achieved without some regulation and tax support; the authors regard their proposal as a "realistic compromise," "disturbing" economic decision-making less than other policies being suggested.[62]

Cognitive Processes

Once again, the perspective is that of bringing some kind of expert advice into the arena of public policy. In this case, the perspective is one of alleging to know the realities of the sphere of business and commerce. The cognitive enterprise is one of assessing how to provide incentives to cut the costs of health care while extending the population being covered with as little disruption of existing businesses and as little growth of government structures and control as possible. As we have seen, the method of justifying this approach is to point to its congruence with American values, and to the power of vested financial interests to oppose radical departures from them. The moral reasons for holding these values as such are not directly addressed.

Enthoven and Kronick do make some explicit provisions for critical and widespread scrutiny of how well their plan, once implemented, actually accomplishes its goals. For one thing, the states that have the money to contract affordable insurance for those presently uninsured are accountable to the consumers who are also voters. Furthermore, Enthoven and Kronick consider it a public good to monitor the health care system and keep the public informed of the data generated. There should be assessments of technology and of outcomes so that providers, sponsors, and consumers will all be well informed.[63] This for the sake, not only of making informed choices, but also of "yielding efficient, high quality care."[64]

But as we noted with respect to Reinhardt's proposal, Enthoven and Kronick do not meet the criteria of the American Public Health Association for improving critical reflection on health care: consumers and poviders should participate in the planning and evaluation; health workers should be trained and educated; and consumers should be educated as to their rights and responsibilities. These concerns are more sharply focused in the proposals put forward by Fein, to which we now turn our attention.

Toward a National Health Policy
Based on Rights to Equity and Cost Control

Rashi Fein has provided a book-length analysis that admirably illuminates why it is that detailed proposals of the kind we have been discussing so far have consistently failed to become public policy in the US.[65] For seventy-five years the debate over NHI has been largely framed by experts, increasingly economists, and carried on by experts, without the necessary mass education and wide participation of the public. Furthermore, the debate has not yet sufficiently clarified what kind of community we are and intend to be. More specifically, the case for health care as a right has not been made, and hence, the case for government obligations for guaranteeing universal access to health care has not yet been made. In this context, Fein amply documents the way in which details of various plans have drawn fire from proponents of alternative detailed plans, always in sufficient numbers to defeat all legislative NHI efforts to date.

In the light of his research and careful reflection, Fein has two very basic suggestions to make. First, instead of making detailed proposals, we should sketch out and debate the principles that should guide a national health care policy. And that debate should not simply be carried on in narrow circles, but in the wide arena of public education, with broad public participation. In addition, since Fein expects neither an immediate consensus nor immediate action, he sets out an agenda for what can and ought to be done while the debate over NHI takes place.

What Justice Demands

Fein's appeal to justice is explicit, and he uses justice in Mill's sense, as implying what someone can claim as a moral right. Health care is a right. There are, for Fein, two parts to this right.

First, there is the right of citizens to have money and resources allocated in ways that square with their perceptions of the benefits of health care and of the benefits that might come from using some of that money and resources in other ways. This right is similar to what we expect from our community "in areas such as national defence, highways, education, fire, and police protection."[66] All of these require budget decisions.

The right to health care is, secondly, a right to an equitable distribution of care, a distribution that "reflects medical need and the costs and benefits of care rather than individual income, wealth, political power, or social status."[67] This right is similar to what we expect "in areas such as access to education, parks, and a basic level of sustenance."[68] All of these require allocation decisions. Fein observes that health care, left to market force, would yield results at variance with the values espoused on behalf of education, the values that move us to offset the consequences of inequalities in incomes. At the same time, equity would not serve us well if the health sector were to be expanded to the detriment of the needs for housing, food, or education.

And so, health care comprises two rights, one to cost control and the other to equity. Both are necessary. Fein contends,

> To achieve them requires a structure that can address both macro and micro health policy, that can determine the citizenry's perceptions and translate them into an effective

program, that enables us to make collective decisions. Such a structure operates at various levels. It is called government.[69]

Fein has another reason for invoking government to address medical care. In the marketplace, people vote with dollars and some have more votes than others. However imperfect, the democratic method can more reasonably be expected to meet the health needs of the total population.

And what tasks do these rights to cost control and equity exact of government? "A universal health insurance program with budget control," says Fein.[70] The government need not produce health care, only concern itself with the amount produced. The government need not operate the delivery system, only hold its agents accountable. The government need not make medical or clinical decisions, only provide incentives and a framework to assist those decisions to reflect medical needs. Achieving equity will increase individual choice insofar as such choices are presently constrained by limited means.

For Fein, to ask what justice demands of us, to ask what we owe one another, is to ask what kind of community we aspire to be in our relations to one another, particularly when some among us are ill or otherwise in need. Competition, as in the free market, should not be the only way in which we relate to one another. Health care should not be allowed to become "just another industry." He says, "it is useful to have parts of our society and economy organized in ways that strengthen our solidarity with others, our charitable instincts, our sense of cooperation."[71] The task of government is to encourage those impulses without which equity cannot be achieved. At the very end of his book, Rashi Fein clearly names those impulses. Whether we will actualize a more equitable system of health care, something we can and should do,

> depends upon whether enough of us care enough to work at translating concepts of decency, humaneness, cooperation, universality, and justice into actions that would protect all

members of the American family. At stake is not only our health care system but the very nature of our society.[72]

Necessity

As Fein sees our current situation, the American public has no sense of a national crisis that would compel immediate and comprehensive action.

> ...America is willing to wring its collective hands about the costs of medical care, but is not yet prepared to act effectively to control them;...it is willing to talk about the need for equity, but not prepared to legislate its enhancement.[73]

However, he does think that NHI will come to be seen as necessary in order to achieve equitable cost containment. This will come about because of increases in price rationing and growing inequities; the middle class will be more and more affected. He gives a detailed account of the forces at work and why they will lead to congressional action on behalf of some form of NHI.[74]

But, if the time for NHI is not yet ripe, what imperatives are there to act now and what actions do they call for? Two kinds of activity should take place. First, since some kind of national, comprehensive, and budgeted system of health care is inevitable, and since detailed schemes keep being defeated, the principles on which NHI can be formulated should now be debated. But this will take time, and there are people who will be suffering if nothing else is done. There are specific changes in the present health care system that will help reduce such suffering. What these should be is guided and rendered imperative by the right to equity and cost containment in health care. What follows is a summary of Fein's agenda for immediate public involvement in efforts to improve our health care system.

1. Unite with others who see the federal budget as a social document expressing our values;
2. question the distribution of expenditures and the tax burden;

3. support the simplification of Medicare's financing, benefit, and payment mechanisms, rejecting means-testing and vouchers for private insurance;
4. reduce benefit disparities in Medicaid;
5. provide long-term care in both programs;
6. extend health insurance coverage;
7. retain funding for public health, nutrition, and other prevention programs;
8. help institutions helping the sickest and neediest;
9. encourage competition, such as HMOs;
10. monitor performance; and
11. become more knowledgeable and more involved, more discerning about pronouncements by professionals, and more able to differentiate analysis and value judgments.[75]

So far as the debate over principles is concerned, suggestions for framing the NHI debate are embodied in discussions of what will succeed and what will facilitate the most informed decision-making.

Chance of Success

Fein argues for health care as a right, but he argues as well that the right to health care is a right both to equity and cost containment. What is more, he believes that one without the other will neither be accepted nor work.

> In the past much of the support for NHI derived from its equity component while much of the opposition was based on the potential for cost increases. Today those who would emphasize equity and those who would emphasize cost control should recognize they need each other. Only if both goals are sought can a sustainable program be erected.[76]

These two goals should be viewed as subect to two constraints: what will be congruent with American experience and attitudes, and what will reasonably mesh with existing ways of financing health care and with the institutions now in place. The current four hundred billion dollar industry is no accident. What is ruled out is a national health service as well as an unbridled free market. A socialized system is too distant from American perspectives; a free market does not take into account what the government now does to support medical care, research, and education.

The reader will recall that all the plans presented in this essay are working within these constraints. However, what this admittedly small sample of proposals is able to illustrate is that the responses to these constraints are far from uniform. Plans 1 and 3, for example, differ sharply with respect to the necessity for retaining private health insurance: Plan 1 abolishes private insurance; Plan 3 very largely has government make use of it. Plan 2 includes suggestions that could move in Plan 1's direction. Fein is much more inclined to leave this matter flexibly in the hands of individual states.

Fein emphasizes the advantages of a state administered, rather than federally administered, health care system. This, he believes, will help make cost containment more of a reality and more subject to the needs and oversight of the voters. This could also result in more attention to calibrating premiums on the basis of ability to pay. Nevertheless, the federal government should be involved to make certain that everyone is insured, that benefits are equal, that mobility across states is taken care of, and that financial assistance to states is provided. Although federal assistance could vary, it should be predetermined, not based on the past expenditures of the state. Obviously, this is to encourage cost containment. Federal assistance is also important to help prevent discrimination and to emphasize prevention with respect to what benefits should be available. Everyone agrees that both preventing illness and catching illnesses in their earlier stages are cost effective as well as humane. Federal assistance should not be an imposed state burden; states are closer to the people for fine tuning, and allocation criteria can be more nearly determined by those who will live with them.

Fein, like Enthoven and Kronick, recognizes how deeply American is the idea of cost sharing. Like the authors of Plans 1 and 2, he is also aware of how unfair it can be. In the end, he believes the success of any plan depends on allowing for some measure of cost sharing. He suggests that unfairness be reduced by the use of tax credits.

Fein agrees with all the other authors that simply mandating employer-provided insurance will not meet the goals of universal and comprehensive health care. Some of the uninsured and underinsured are not predictably cared for in this way.

Cognitive Processes

As the reader will have observed, a constant and consistent theme in Fein's proposals is that of participation by the public. He favors a major role for state governments largely because they are much more subject to knowledgeable input and control by citizens and their organizations. Although he does not specify how to accomplish it, he urges that the debate over NHI not be left to various experts, but that it be expanded to include the public much more, a public that should be provided a greater degree of knowledge about what is at stake. Here he is resonating to similar principles espoused by the APHA mentioned previously. All three of the other proposals discussed above can be faulted for giving insufficient, explicit attention to these principles. He is concerned to empower individuals within a heightened democratic process and by means of a benefit package that poses no barrier to preventing illnesses and treatment of illnesses in their earliest stages. These principles are also shared by the APHA.[77]

It should not be overlooked that, from Fein's perspective, the underlying question posed by a debate over health care is the question of what kind of community we are and ought to be. That question itself invites wide public participation in all aspects of health care, the conceptualization of it as well.

A Concluding Assessment

One purpose of our essay was to gain some understanding as to why it is that the US is alone among major industrial countries without comprehensive and universal health care coverage for all of its residents and citizens. For that purpose, we analyzed four proposals, taking up some recent significant proposals not specifically analyzed in Fein's own appraisal of the health care debate. With the use of four criteria, it was possible to see some of the ways in which those agreed on the need for NHI nevertheless have reasons to criticize one another's plans, and also to work against plans other than their own. What we have found lends credence to Fein's plea to explore what principles should guide the debate and the proposals, laying aside details for which arguments are not even usually provided.

But are any of the detailed proposals for NHI convincing or worthy of further defense? Should the US adopt some form of NHI?

Necessity

We chose to examine four proposals that seek to change the present American health care system. All four of these proposals have significantly narrowed the options. They see no prospect for moving toward a totally free market system, or toward a national health service of the sort found in Britain. Furthermore, all of them have a role for the federal government to assure universal health care access and cost containment. But as we have observed in our analysis so far, they do differ as to the nature and extent of federal involvement in the health care system. What is behind these differences will have to be clarified before negotiations over them can move toward a plan sufficiently persuasive to be implemented. We will make some suggestions below.

There are at least two other issues not sufficiently addressed by the authors we have examined. The first is implicit in Fein's observation that the present health care system is not perceived as a crisis, because the great majority of Americans have access to health care of a high quality. But, as we shall argue later, the fact that individuals are now dying and suffering to a degree and in ways which are quite preventable in this wealthy country, is a severe crisis. To take that view is to propose a theory of justice not clearly found in the proponents of NHI we have been discussing.

There is another issue raised by the commitment to cost control in all of these proposals: controlling costs poses a threat both to the quality of care available and to the realization of equal access NHI is supposed to assure.

Chance of Success

The fear that NHI will compromise the high quality of health care in the US is one that Reinhardt regards as realistic. Other countries with universal access to health care guaranteed by their national governments cannot match the US in quality of care. Although Reinhardt does not make this point, it can be argued that the quest for excelling in medical care is at least one major reason that the US leads the other industrialized nations in the percentage of its GNP devoted to health expenditures. It is possible to see this as an advantage rather than a liability, one that the US should not necessarily or readily relinquish. Indeed, one could also argue that other countries should endeavor to increase the percentage of their GNP spent on health care.

The reason for spending more for health care is not confined to the question of quality. Goverment control over health care costs may take the form of creating inequalities in the care available and offered. In Britain, for example, certain services are not made available to individuals beyond a certain age.[78] In Russia, quality care is not available for the great majority, and that same majority must wait in line for health care as they do for so many other services.[79] In Canada, prospectively limited, global budgets are beginning to lead to the curtailment of services and increased waiting periods for the care available.[80] Reinhardt is rather pessimistic about any prospect in the US that the majority now enjoying access to health care are willing to pay the price to grant others access to that same high quality of care. His solution is to accept, at least for the foreseeable future, a two-tiered system that is expressly less than equitable. Others, like Beauchamp and Childress, speak in a similar vein, of guaranteeing everyone "a decent minimum" in health care, without trying to specify what that would be.[81] Still, others are advocating the rationing of services, singling out certain groups for receiving less care than what is provided for others.[82] Given the evidence that insufficient and unequally distributed care occurs in countries with nationally comprehensive and universally accessible health care systems, and given the increasing proclivity to justify inequalities and reduced expenditures for care, those in the US who currently have health insurance have some grounds to resist change; they have grounds also to oppose proposals for NHI that increase the federal role in the health care system while requiring the federal government to cut the costs of that system. It is not surprising that Fein is worried that the federal government may not be willing to spend enough money to realize the goal of NHI, especially at a time when there is such pressure to reduce the federal deficit.[83] The case for NHI will not be made, and these fears will not be allayed, unless the adoption of NHI bids fairly to decrease, not increase, discrimination, and unless NHI bids fairly to increase, not decrease, the quality of health care. The great majority of Americans are now insured and cannot be expected to vote for less health protection than they now have. At the same time, some propose NHI because they believe that the health care system in place will create hardships for the majority of people down the road without some form of NHI. This is supposed to happen because of uncontrolled health care costs, the cure for which is to bring them under control. But, as we now wish to argue, unless this is done in a way that protects and enhances basic human rights, the alleged cure may be worse than the alleged disease.

The Demands of Justice

We agree with Fein: what we do about health care reflects what kind of community we are; and what we think ought to be done about health care reflects what kind of community we think we ought to be.

To begin with, then, there are certain human relations that are requisite for forming and sustaining communities as well as its individual members. An obvious one is that human life requires protection. Such protection has no less than two facets: prohibitions and sanctions against killing, and care and nurture, especially in those times for each of us when our lives depend on it. No less obvious a requisite of community is procreation itself, and the care and nurture that assures that there will continue to be individuals for forming community. These requisites of community are themselves not simply individual acts, but cooperative relations that are part of what we mean by speaking of a community.

The spawning and protection of human lives is not something individuals can accomplish by themselves. This is especially true when an individual becomes ill or disabled. In an industrialized society with its professional kinds of services, help is something requiring compensation. To the extent it is not compensated by the community, individuals will have to buy such services. If they do not have sufficient money, with or without a job, their lives may quite simply be on the line, so also will be the lives of any dependents. All of this seems too elementary to say. Consider, however, that a parent may be charged with child abuse for failing to seek medical care, but that same parent in the US has no strict right to expect that care to be given or paid for out of communal resources should that be required for the care needed.

What we are proposing is that the kind of communal protection of lives provided through armed services, police forces, and fire departments be extended to health care. These structures of communal concern for the protection of the lives of individuals are needed for the same reason that communal health care provision is needed; no one can adequately secure his or her life on his or her own. Lives lost because health services cannot be bought and are not communally assured are no less a responsibility of shared, communal living, and no less precious than those that would be lost if we had no armed services, no police force, or no fire departments, voluntary or compensated. What can be claimed as rights and affirmed as communal responsibilities in any one of these services can be similarly claimed and affirmed in all of them. This is how we sustain individual lives; this is how we sustain communities. To do these things is to serve the fundamental right to life and, we would add, to the nurture and care necessary to life, individually and communally.

Fein made the case for budgeting all of these services. But, whereas he focused on the necessity of this for cost control, we wish to emphasize budgeting for the sake of eliciting the necessary funds to assure the existence and adequacy of these life protecting services. And, instead of speaking of containing costs, the desire to protect human rights moves us to speak instead of responsible, honest, accountable, and nonwasteful expenditure of the money allocated by and for the community.

But what about the questions being raised about rising costs, and the possibility that vital services and goods other than health care will shrink unduly unless health care costs are frozen or reduced? Such problems should indeed be addressed. However, if we take as our primary responsibility that health policy aim at preventing preventable deaths, this legitimate concern about how money is allocated should, at least for the present, be addressed very differently.

To begin with, the tendency is to view the limits to be placed on health care as a tradeoff among a variety of public goods and services, all of them vital and worthy. However, Americans are willing to make sacrifices when they see their individual and collective lives at stake. They do so by transfers of wealth from one sphere to another. Consider the enormous transfer of money for national defense throughout the cold war; research and education are also boosted whenever they are assimilated to national defense as they sometimes are. Consider also how some of our money is spent in the so-called private sector. We refer to the private sector as "so-called" because all of the money in it is money obtained from others in both the private and public sectors of various communities, our own included. We spend enormous amounts for recreation, for the salaries of entertainers and executives, for goods we can never use because they are so numerous, and for products, such as drugs, alcohol, and tobacco to name a few, injurious to our health, which directly drive up the amount of illness and the costs of care. Those who feel an obligation to protect human lives would willingly return a larger share of what they have to undergird health care if they had reasonable assurances that the money would be responsibly spent. To the extent that the public sector mismanages money, responsible individuals will rightly oppose transfers from the private to the public sector. Containing costs by itself does not work on behalf of human rights. Cutting waste and inefficiency without cutting services is quite another matter.

With the focus on protecting lives, there are other areas that should be the first target for cutting costs. These are expenditures that drive up health care costs without directly increasing the quality and quantity of health care. Consider the large outlays for litigation in the health sphere. And consider what accompanies these outlays: high prices for malpractice insurance, and expensive, unnecessary procedures in the practice of defensive medicine, understandable, though regrettable, under the circumstances. Then, too, there is a tendency to seek excessive profits for therapeutic agents; AZT is a case in point. The case of AZT also underlines what we wish to recommend. Namely, that these various inflationary pressures on medicine require public scrutiny to bring them under some kind of control. This, we contend, would be justifiable cost containment. These are moves to make before it is seemly to advocate rationing, particularly rationing that can only be characterized as discriminatory, such as excluding older persons from care solely on the basis of age.[84]

Cognitive Processes

There is not enough space to expand on the arguments for our particular view of justice and its implications for NHI. However, something should be said about the context in which such arguments, and the responses to them, are justly carried on. How health care is to be provided and how its quality is to be maintained and enhanced is not for experts to decide. Power exercised by a few, with little or no accountability to all who are affected by the policy in question, violates the basic presumption of what is requisite to cooperative communal activities, namely, mutual responsibility and accountability to one another as moral equals. Justice includes the right and responsibility to participate in communal decisions. Democratic structures that allow for accountability to a voting public, and protection of individuals and minorities against the possible violation of human rights by majority rule, should be explicit concerns of any NHI policy.

If one compares the provision of health care with the provision of education, as Fein does, there is more to it than the use of public money to help equalize opportunities. Education helps to empower people and increase their freedom. At the same time, it does not predictably do so without strict accountability to the people it serves, and without the participation of all, including those who have no one to send to school. Local control is a significant principle of democracy stressed by Fein and too little stressed by NHI proposals. From the point-of-view of democratic justice and monitoring of the health care sector, people have good reason to oppose NHI plans that insufficiently empower and involve the persons who are to be served. And it should not be overlooked that education benefits from having both private and public education; so does freedom of association, a hallmark of American democracy and essential to its health. It may be that a

mix of private and public financing for health care is not so much
a constraint on health policy, but a justifiable way of enhancing
freedom and competent health care.

It should be clear that there are many legitimate and reasonable bases for opposing NHI in the US. It should be equally clear
that with the proper focus on protecting human life in nondis-
criminatory ways, and the proper democratic safeguards, some
form of NHI can be justly instituted in the United States.

Notes and References

[1]Rashi Fein (1989) *Medical Care Medical Costs: The Search for
a Health Insurance Policy.* Harvard University Press, Cambridge, MA,
p. 34.

[2]*Ibid.*

[3]*Ibid.*, p. v.

[4]Larry R. Churchill (1987) *Rationing Health Care in America:
Perceptions on Principles of Justice.* Notre Dame University Press, Notre
Dame, IN.

[5]For competing theories, *see* the chapter on Justice in Tom L.
Beauchamp and James F. Childress, *Princlples of Biomedical Ethics.*
Oxford University Press, New York, NY, pp. 256–306.

[6]Rashi Fein, *op. cit.*, 198.

[7]Beauchamp and Childress, *op. cit.*, 257.

[8]John Stuart Mill (1957) *Utilitarianism.* Bobbs-Merrill, New York,
NY, p. 62.

[9]*Ibid.*

[10]David Himmelstein and Steffie Woolhandler (1989) "A National
Health Program for the United States: A Physician's Proposal," *New
England Journal of Medicine* **320,** 102–108.

[11]*Ibid.,* 102 .

[12]Gene Outka (1974) "Social Justice and Equal Access to Health
Care," *Journal of Religious Ethics* **2,** 11–32.

[13]Himmelstein and Woolhandler, *op. cit.*, 102.

[14]Larry Churchill, *op. cit.*

[15]Himmelstein and Woolhandler, *op. cit.*, 105.

[16]*Ibid.*, 102.
[17]*Ibid.*
[18]*Ibid.*, 103.
[19]*Ibid.*
[20]*Ibid.*, 107.
[21]*Ibid.*
[22]*Ibid.*, 102.
[23]*Ibid.*, 103.
[24]Uwe E. Reinhardt (1987) "Health Insurance for the Nation's Poor," *Health Affairs* Spring 1987, 101–112.
[25]*Ibid.*, 108–109.
[26]*Ibid.*, 102.
[27]*Ibid.*,
[28]*Ibid.*, 103.
[29]*Ibid.*,
[30]*Ibid.*, 103–104.
[31]*Ibid.*, 110–111.
[32]*Ibid.*, 102.
[33]*Ibid.*, 104.
[34]*Ibid.*
[35]*Ibid.*, 103.
[36]*Ibid.*
[37]*Ibid.*
[38]*Ibid.*
[39]*Ibid.*, 104.
[40]*Ibid.*
[41]*Ibid.*, 105.
[42]*Ibid.*
[43] Edward Kennedy and Henry Waxman, S768 and HR1845.
[44]The proposal by Enthoven and Kronick, referenced at note 50, has this as one important feature.
[45]Reinhardt, *op. cit.*, 107.
[46]*Ibid.*, 110–111.
[47]*Ibid.*, 111.
[48]*Ibid.*, 110.
[49]*The Nation's Health* 1990 (Mar), 11.
[50]Alain Enthoven and Richard Kronick (1989) "A Consumer-Choice Health Plan for the 1990's: Universal Health Insurance in a

System Designed to Promote Quality and Economy." (First of two parts) *New England Journal of Medicine* **320,** 29-37; (Second of Two Parts) **320,** 94-101.

[51]*Ibid.*, 31.
[52]*Ibid.*, 29.
[53]*Ibid.*
[54]*Ibid.*, 30.
[55]*Ibid.*
[56]*Ibid.*, 31.
[57]*Ibid.*, 30.
[58]*Ibid.*, 31.
[59]*Ibid.*, 101.
[60]Chapter 23 of the Acts of 1988 signed into law by the Governor, April 21, 1988.
[61]Enthoven and Kronick, *op. cit.*, 97.
[62]*Ibid.*
[63]*Ibid.*, 35.
[64]*Ibid.*
[65]Rashi Fein, *op. cit.*
[66]*Ibid.*, 194.
[67]*Ibid.*, 194–195.
[68]*Ibid.*, 195.
[69]*Ib.d.*
[70]*Ibid.*, 196.
[71]*Ibid.*, 192.
[72]*Ibid.*, 222.
[73]*Ibid.*, 216.
[74]*Ibid.*, 216–220.
[75]*Ibid.*, 220–222.
[76]*Ibid.*, 215.
[77]*The Nation's Health, op. cit.*, 11.
[78]Henry Aaron and William B. Schwartz (1990) "Rationing Health Care: The Choice Before Us," *Science* **247,** 416–422.
[79]Daniel S. Schultz and Michael P. Rafferty (1990) "Soviet Health Care and Perestroika," *American Journal of Public Health* **80,** 193–197.
[80]B. Brown (1989) "How Canada's Health System Works," *Business & Health* July 1989, 28–30; Adam L. Linton (1990) "The

Canadian Health Care System: A Canadian Physician's Perspective," *New England Journal of Medicine* **322,** 197–199.

[81]Beauchamp and Childress, *op. cit.*

[82]Daniel Callahan (1987) *Setting Limits: Medical Goals in an Aging Society.* Simon and Schuster, New York, NY; Norman Daniels (1988) *Am I My Parents' Keeper?* Oxford University Press, New York, NY.

[83]Personal communication.

[84]Callahan and Daniels, *op. cit.*

How Just Must We Be?

Leonard M. Fleck

Introduction

Over the past two years the debate has reemerged over whether or not the US ought to replace its current patchwork system for financing access to health care with some system of national health insurance.[1] This debate last occurred in the mid-1970s, but it was effectively squelched by a worsening economy and escalating health care costs that made national health insurance seem unaffordable. Growth of more than 25% in the ranks of those without any health insurance during the decade of the eighties has renewed this debate. Currently, there are about 37 million Americans without any health insurance. Most are poor, but employed—70% of them are employed more than twenty hours per week. But these are generally low-paying service sector jobs for small employers who cannot afford to provide health benefits.

Since most of these individuals are young and healthy, lack of health insurance has no practical consequence. Only 10–20% of these individuals are likely to need medical or hospital care in any given year. Ten years ago, these individuals could have been relatively confident that they would get the care they needed through the charitable budgets of hospitals or charity care given by physicians. But during the eighties, intense pressures for cost containment have eroded the capacity of hospitals to provide

From: *Biomedical Ethics Reviews • 1990*
Eds.: J. Humber & R. Almeder ©1991 The Humana Press Inc., Clifton, NJ

such care. A tragic illustration of this point is provided in testimony from an emergency room physician in Detroit. He told of
a young woman in her thirties, brought to the emergency room
unconscious from a head injury sustained in a mugging and rob-
bery. She was dressed in jeans and had no health insurance card.
It took seventeen hours to find a neurosurgeon who was willing
to do the necessary surgery without any promise of payment, by
which time her condition had deteriorated substantially. She died
three days later, at which time it was discovered that she did have
Blue Cross/Blue Shield insurance.[2]

How should we judge this outcome from a moral point of
view? Is it morally acceptable to conclude that this was terribly
unfortunate, and no more? Or was this unjust? More generally,
since a similar outcome is possible for anyone who has a serious
health problem and lacks health insurance, do we as a society
treat all the uninsured unjustly by failing to provide them with
secure access to needed health care? As noted earlier, we do
speak of "health benefits" and "charity care," which would suggest that there are no moral entitlements rooted in justice to such
care. But I shall argue in this essay that such language is mistaken, and that national health insurance today is something required of our society as a matter of justice. This last comment
does not imply that this is a conclusion that can be neatly deduced
from some favored theory of justice. It cannot. Rather, as I will
show, this conclusion will follow as the result of balancing an
array of moral and nonmoral considerations. In that respect it will
be a product of a certain reflective equilibrium.[3]

I will not defend some generic form of national health insurance. There are a number of proposals for national health insurance on the policy agenda today that, in my judgment, are not "just
enough."[4] It would be too tedious to attempt to do a detailed
evaluation of each of these proposals. Fortunately, these proposals
have a number of common features that, although their proponents see as their strengths, I shall argue represent their greatest

deficiencies from the perspectives of both equity and efficiency. These are the features: first, they are all committed to a pluralistic approach to financing health care. Currently, there are about 2,000 health insurance and HMO-type plans for health financing in America, each of which offers a different package of services at different prices. What is rejected by all these proposals is the idea of a single mechanism for financing health care, such as we would find in Canada or Great Britain.

Second, all these proposals are committed to a mixed public/private system for financing health care. That is, the government would be responsible for funding or subsidizing access to health care for those who could not afford health insurance themselves, whereas the private sector would sell health insurance to companies and individuals who could afford to pay. What is clearly rejected is an entirely public mechanism for financing access to health care.

Third, all these proposals are committed to a multitiered health care system—not everyone would be guaranteed the same package of health care services. Put crudely, there would be Good Will care for the poor, K-Mart care for the working poor and near poor, Sears care for the bulk of the working class and middle class, and Bloomingdale care for the upper classes. What is rejected is the idea that equity requires that all in our society have access to the same package of health services.

Fourth, all these proposals are committed to guaranteeing only minimally decent care for all in our society. It is only in this somewhat limited sense that everyone in our society would be assured access to the health care system. What is usually rejected on the basis of alleged cost considerations is the idea that everyone should have access to a very comprehensive package of health services, such as is available to Canadian citizens.

Fifth, all of these proposals recognize the need for containing escalating health care costs, and they share the belief that this is best achieved through market competition among organized groups of providers/financiers of health care. This represents a rejection of the use of government regulations as the primary mechanism for achieving cost containment objectives or other efficiencies in the delivery/financing of health services.

Finally, these proposals generally share the view that resource allocation decisions ought to be made by individual consumers (in the light of their personal budgets for health care) or by organized groups of providers at the local level. This would mean that resource allocation decisions would be essentially private affairs beyond the purview of public scrutiny. This represents a rejection of centralized decision making on resource allocation by government bureaucrats.

There is always the risk of criticizing a "straw man" if the generic description outlined above is the focal point of discussion. Rather than take that risk, it is my judgment that the Enthoven-Kronick proposal represents the best of the lot, all things considered, but taking special account of matters of justice. Nevertheless, it is also my judgment that it is not "just enough." This is not a judgment that I offer from some idealized moral and economic perspective. As I will explain, the source of this judgment is a nonideal framework of justice.

There is no "perfect" policy for universal health care financing. However, there is at least one feasible policy proposal that is rationally preferable from a variety of value perspectives. Specifically, the alternative proposal that I would endorse is that offered by Himmelstein and Woolhandler.[5] Their proposal mirrors closely the Canadian system for financing health care, which means it has the practical advantage of being more than an interesting theory. It has been in place for twenty years in a society that is close to our own, culturally speaking, which means there is enormous relevant data available for assessing its strengths and weaknesses. Its primary features are the opposite of the six features outlined above in competing proposals. That is, it is a very comprehensive, single health plan available to all Canadian citizens. The government is the single source for financing the cost of all health services, and it is the same package of health services that are available to all Canadian citizens, no matter what their socio-economic status might be. These features say a lot about the basic equity of this plan, but more interesting is the fact that the Canadians have been very successful thus far in controlling escalating health care costs.

Robert Evans, a Canadian economist, notes that in 1970 (when Canada introduced its system of national health insurance) both Canada and the US spent 7.4% of their GNP on health care. Since then, Canadian health expenditures have risen moderately to 8.7% of GNP while expenditures in the US have jumped to 11.5% of GNP.[6] These statistics require some interpretation, which is offered below. Note that these results were achieved without sacrificing quality of care in Canada. Further, Canada achieved effective cost control in health care by committing itself to universal access and by employing a single mechanism for financing their health care system. This is contrary to the received wisdom in the US, which says that we cannot afford to provide health care for all. The lesson from Canada is that we cannot afford the inequitable system we currently have.

The plan for the remainder of this essay will be as follows. First, I shall offer a brief description and explanation of the framework of analysis for this essay, which I refer to as a nonideal framework of justice. Second, in order to engage this framework I will show that our current methods for financing health care in America are both inefficient and inequitable. Our current health financing mechanisms violate very widely agreed on considered judgments of justice, which are not too closely tied to controversial philosophic theories of justice or philosophic theories of health care justice, such as those advocated by Veatch,[7] Engelhardt,[8] or Daniels.[9] Further, I also have to be able to show that patchwork reform (better known among political scientists as disjointed incrementalism), the standard mechanism for addressing public policy problems in the US, is also inadequte from the perspectives of both equity and efficiency. More positively, I must show that there are powerful moral and economic considerations that require the relatively radical reform of our health care system that is represented by national health insurance.

Third, political support is building in our society for some form of national health insurance. The argument of the American Medical Association[10] and a number of policy analysts[11] is that the only proposals that are really politically feasible in our society are those that have the six features listed above, because those features define the deep moral, political, and economic values inherent in our history and political culture. That concatenation of values represents a libertarian conception of justice. If these were the necessary and sufficient criteria for assessing competing proposals for financing health care, then the Enthoven-Kronick proposal would likely best satisfy these criteria. But the analysis in the second part of the paper will have undercut greatly both the moral and economic legitimacy of these criteria. Consequently, I will propose instead an alternate set of thirteen criteria that will provide a more comprehensive and appropriate basis for making a comparative assessment of the Enthoven-Kronick and Himmelstein-Woolhandler proposals.

Nonideal Justice: Some Methodological Remarks

From the perspective of some ideal theory of justice, all feasible proposals for national health insurance are unjust. Philosophers might be able to sleep comfortably at night with this conclusion because they would not have allowed their standards of justice to be compromised by concerns related to political or economic feasibility. However, morally sensitive public policy-makers could hardly sleep with their consciences if they had to believe that all their policy choices were unjust, the practical implication being that it did not matter, morally speaking, which particular policy option they pushed. From this perspective moral progress in the making of public policy would make no sense. I wish to argue, however, that this is not a conclusion we should accept. John Rawls has written that "in practice we must usually choose between several unjust, or second best arrangements, and then we look to nonideal theory to find the least unjust scheme."[12] But Rawls himself provides us with little guidance on nonideal theory beyond speaking of the need to achieve a "balance of imperfections, an adjustment of compensating injustices."[13]

There are two key insights that I have culled from these remarks by Rawls. First, when we are talking about justice "in practice," we are always talking about some very specific context. I refer to this as a "problem/decision sphere." Second, the implication of being able to talk about a "least unjust scheme" is that it is logically and morally appropriate to speak about possible states of affairs (policy options) as being more or less just relative to one another. I refer to this as the possiblity of "moral meliorism." These are starting points for articulating a framework of nonideal justice, only the bare bones of which I can outline here.[14]

Defining Our Problem/Decision Sphere

The notion of a "problem/decision sphere" is a pragmatic, highly relativized notion. It is defined by what is manageable in the way of solving a problem and effecting change, given a host of political, economic, organizational, normative, and technological constraints peculiar to that problem. The precise boundaries of this sphere are subject to some shifting as the relevant political actors negotiate the definition of a problem. With reference to our topic, "the" problem that has served as a focal point is the growing number of those without any health insurance.

As we shall see, the definite article needs the scare quotes because we are really talking about an amalgam of problems that might be loosely grouped under the phrase "financing health care in America." Some people are uninsured because they are unemployed. Others are employed but uninsured because their employers do not provide group health insurance and they cannot afford the cost of purchasing a private insurance policy. Still others may have insurance for themselves as workers, but not be able to afford insurance to cover their spouse and children. Divorce may leave a wife and children without health insurance. Some who could afford private insurance will forego it as a gamble because they would prefer to use the money for other things.

Others are told by insurance companies that they are uninsurable because they are very poor insurance risks, such as those who are HIV-infected, or who have the wrong sorts of family health histories. Others who are far below the poverty line are not eligible for Medicaid in their states because income eligibility requirements have been set low to reduce costs to Medicaid in those states. Finally, there are millions of individuals whose health insurance would be grossly inadequate if they were afflicted with an illness that required anything more than a short hospital stay.

In the past, most of these people received needed health care on a charity basis. The care they received was not that expensive, relatively speaking. But with all sorts of technological advances in medicine, there are many more things that can be done for most medical problems, usually at greater cost. This has fueled the problem of escalating health care costs, but it has made especially burdensome the provision of charity care to the uninsured.[15] Of course, relatively little "true" charity care has been provided. Most of the time, physicians and hospitals have shifted the costs of providing such care to other third party payors.

Medicare and Medicaid have been among those third party payors. However, as the taxpayer revolt of the late 1970s spread, intense pressure to hold down public budgets resulted in Medicare and Medicaid paying hospitals and physicians deeply discounted fees for their services. That meant more costs had to be shifted to private third party payors, especially large corporations. These corporations saw the health care system hijacking their profits. Further, they contended they were being treated unfairly because they were effectively paying for health benefits for millions of small businesses that failed to provide health benefits to their employees. Small businesmen, in turn, complained that they cannot afford the insurance premiums which, for their employees, would be substantially higher than the premiums paid for comparable policies by large corporations.

One upshot of this analysis is that those with the greatest health needs are least likely to have access to the health care system that could meet those needs. This will strike many as being morally indecent, something that is just not compatible with a society that claims to be caring and compassionate. Beyond that, our collective moral intuitions are somewhat confused. After all, health care insurance is a "benefit" associated with having a good job; but then we do have very effective health technologies that often make the difference between life and death, and very few Americans are willing to tolerate the denial of such technologies to those unfortunate enough not to have health insurance. Still, there is a problem with moral distancing here, for those average Americans, like average corporate benefits managers and average bureaucrats, all assume that the responsibility for providing that health insurance and access to the health care system belongs to someone else. Thus, all those who have engaged in cost shifting or patient dumping see themselves as acting in a morally legitimate way because they are merely protecting themselves from being unfairly taken advantage of. This rationale, of course, fails to take into account the plight of patients with serious health needs who are victimized by these defensive moves.

We now have a fairly crude description of our problem/ decision sphere. We reserve for the next section a more precise moral analysis. What we wish to emphasize from a methodological perspective is the highly localized nature of this problem/decision sphere. There may be serious injustices associated with our health care system as a whole, or with our whole sytem of social welfare, or with capitalism as our economic system. But the point of nonideal moral analysis is that these larger injustices can be legitimately ignored for purposes of achieving some incremental moral improvement in some limited area of public policy.

What I would define as our problem sphere is this: what policies ought we adopt for financing health care that will assure equitable access to costworthy health care for all in our society? Many policy analysts would prefer a more restricted definition of the problem, such as this: what policies ought we adopt for assuring that the uninsured have access to our health care system? As I will argue below, this more restricted problem definition will only yield greater inefficiencies and greater inequities in our health care system.

The basic difficulty is that the uninsured are a very heterogeneous group (health-wise and income-wise), and the group is very fluid, due mostly to shifting employment opportunities. This suggests the need for several sorts of programs that would be funded by some mix of private and public sources, a mix that has proved highly unstable in the recent past as escalating health costs have precipitated defensive strategies for shifting these costs elsewhere. This has created what I have referred to elsewhere as "moral twilight zones," zones of moral responsibility in which the distinction between unfortunate outcomes and unjust outcomes is badly blurred.[16] The case of the Detroit woman mentioned earlier falls squarely into one of those moral twilight zones. Such moral twilight zones are open to legitimate moral criticism just because they do blur so badly the identification and assignment of moral responsibility. Our current patchwork system for financing health care contributes substantially to the creation of such moral twilight zones. Hence, proposals for "reforming" our system for financing health care that only add more patches to the system ought to be viewed presumptively with moral suspicion.

These remarks hardly constitute ironclad evidence of the adequacy or correctness of my delineation of our problem/decision sphere, but such evidence is not needed at this point. Our method of inquiry is very much in the spirit of John Dewey. Later stages in the inquiry will reveal whether our initial problem formulation is adequate enough.

Moral Meliorism and Nonideal Justice

The second key element in our framework of nonideal justice is that of moral meliorism. The basic idea here is that we have done something morally commendable if, in our problem/decision sphere, we can bring about a more just state of affairs, all things considered, than what existed previously, even though this improved state of affairs will still fall short of what an idealized conception of justice might require. This gives us an internal reference point to measure socio-moral progress, which is of great practical advantage when our objective is to bring about a more just state of affairs in a nonideal world. External reference points risk the charge of either irrelevance or utopianism, both of which undercut the possibilities for real (but nonideal) moral progress.

The alternative risk is that of complacency or moral minimalism. Norman Daniels has noted that isolated proposals for reform may make it more difficult to rectify deeper injustices in a social system.[17] This may be true in very specific circumstances, and so, this is a perspective that will sometimes have to be considered. But as a more general objection to meliorism, this point fails because it pushes us in the direction of moral utopianism—rejecting smaller real reforms that are possible in favor of idealized states of affairs that are remotely possible, at best. Meliorism is rooted in the reasonable empirical claim that, in democratic societies, political actors have only limited power to shape events and limited opportunities to do so. This has the virtue of protect-

ing social stability, which is an important background condition
for preserving justice in transactions over long periods of time.

Still, there is the legitimate concern, expressed by Daniels,
that meliorism would degenerate into the moral myopia of
bureaucrats who would reduce the morally desirable to the politically feasible. As a practical matter, the best insurance against
this happening would be strong practices of open, public moral
criticism and debate built into the workings of key social institutions. Such debate can do much to expand the range of what is
feasible. Still, if we take moral meliorism seriously, then we cannot
ignore feasibility factors. Otherwise, pious posturing becomes a
substitute for effective moral reform.

Financing Health Care in America: A Critical Assessment

We argue here that there are compelling moral reasons for
reforming, in significant ways, the financing of health care in
America. Before embarking on that critique proper, there are two
prior questions that need to be answered. First, why is health care
"special," morally speaking? That is, why should we think that
access to health care should be governed by some set of moral
considerations? Second, why should health care be thought of as
a public good rather than just another private consumer good to
be distributed in accord with ability to pay?

Health Care as a Morally Significant Good

Health care is special because it is intimately connected with
two basic moral values, namely, the preservation of human life
and equality of opportunity. To be sure, this is a contingent rather
than an essential fact about health care in our society. If the year
were 1900 instead of 1990, then having access to either a physician or a hospital would mean that one had no more than a 50/50
chance of getting better rather than getting worse as a result of

access to that health care system. Most physicians at that time were poorly trained. Hospitals were poorly equipped for the most part. There was not what we would understand as scientific medicine today.[18] A right to health care could hardly have been thought of as a blessing then.

Today, however, we have effective scientific medicine. Millions of lives are saved each year now that would have been lost if we had only the medicine of 1900. Further, enormous amounts of suffering are eliminated or ameliorated through contemporary medicine, including substantial recent breakthroughs in rehabilitative medicine. We concede that such medicine has often been used too aggressively and inappropriately in terminal situations, and the burden of chronic illness in our society has been increased and prolonged because our medical successes have often taken the form of halfway technologies rather than curative interventions. Still, virtually no one would doubt that having access to our contemporary health care system represents a substantial net benefit. This reinforces our main point: a health care system that has the capacity to determine whether one lives or dies at a certain point in time ought to be an object of serious moral concern.

There is another way in which we can make this last point more directly. We used to think that death was mostly a "natural event," something that happened to individuals as a result of natural causes, such as the progression of disease. Under such circumstances, no one was responsible for the death of an individual. But for most people today, death will not be a product of pure natural events. Rather, a large number of personal and social choices will determine the time and circumstances of one's death, which is to say that moral responsibility will be assignable. But it is not just doctors and patients who make these choice. Choices are made at the social level that determine who lives and who dies. The single most important of these factors will be whether or not someone has health insurance, for that determines more than anything else whether one has access to our health care system.

Most needed health care is not a matter of life and death. That, however, does not diminish the moral significance of that needed health care. For the other major moral value connected with scientific medicine is equality of opportunity. Our society is not committed to a strong egalitarianism when it comes to the distribution of income or wealth. From within our nonideal moral framework, this is something we can accept readily enough. We are, however, committed to the notion of equality of opportunity. Rawls offers, I believe, a reasonable interpretation of our collective moral intuitions in this matter. We are better off in terms of social welfare if we permit some degree of inequality in wealth and income. But we protect the integrity of our more basic moral intuitions, that all are entitled to equal concern and respect, if there is effective equality of opportunity to access these more desirable social positions. It seems this is why we provide free public education for twelve years and highly subsidized education thereafter. It is a contingent fact, but a fact nevertheless, that we have a knowledge-based society. If many of our citizens were denied access to education through, for example, financial barriers, then we would have effectively denied them fair equality of opportunity.

Norman Daniels has cogently argued (in my judgment) that much of health care is very closely connected to effective equality of opportunity, just like education.[19] Disease and illness often interfere significantly with our capacity to carry out most of our life plans. Imagine a youngster who is afflicted with a severe form of anemia, that makes it very difficult for him to learn in school. There is a drug that can alleviate this anemia to a large extent, but it is very expensive. His parents cannot afford it because they do not have health insurance. Could we honestly say that our society has guaranteed equality of opportunity to this youngster? Certainly, it would be unfortunate if there were no medical intervention that could relieve this anemia, the result being that this youngster would go through life significantly handicapped. But when the relief is there, and it is simply denied to that youngster because his parents cannot afford that needed medical care, then this is not simply unfortunate. Rather, the more fitting description is that it is unfair because this youngster is denied equality of opportunity, not just in some very limited aspect of life (like being able to play the piano), but for virtually any life plan that he would choose.

A third reason for thinking that access to health care is a matter of special moral concern is suggested by the President's Commission. They note that we are all vulnerable to disease and death. Given this, health care has special moral significance because "it expresses and nurtures bonds of empathy and compassion. The depth of a society's concern about health care can be seen as a measure of its sense of solidarity in the face of suffering and death."[20] An apt analogy would be society's response to a massive natural disaster, such as an earthquake. Those who were spared the devastation could hardly expect that they could be indifferent to the plight of the victims and escape moral criticism. Under such circumstances, there is a strong moral obligation to provide some level of assistance. It would seem that some line of thinking like this motivated the conclusion of the Commission that "society has an ethical obligation to ensure equitable access to health care for all."[21] The precise nature of this moral obligation is somewhat ambiguous in this report. As I have suggested elsewhere, the Commission seems to think of this obligation as being more a matter of beneficence than justice.[22] For our present purposes, we can avoid that debate. The important conclusion is that access to health care is correctly thought of as a matter of moral concern.

Health Care as a Public Good

To justify public policy governing the distribution of health care, we must show that it is legitimately thought of as a public good rather than a private consumer good. Space permits presenting only some persuasive considerations for this view.

First, if access to health care is connected with fair equality of opportunity, it is also properly thought of as a public interest. How could equality of opportunity be preserved if it were left to private actors governed by their private interests? A purely private system of education would virtually guarantee that the poor and their children would be permanent members of an underclass.

Second, though we often speak of our health care system as being private (because it is not owned by the government for the most part), huge public investments have been made in the system. For the most part, public dollars have built the hospitals, trained the health care professionals to an exquisitely refined degree, and underwritten the costs of the research that have yielded the cornucopia of medical advances we all take for granted. It is safe to say that we would have a substantially less sophisticated health care system, and the vast majority of us would be a lot less well off if those public investments had not been made. Those public investments, of course, are taxes, which have been paid by the vast majority of the uninsured who, after all, are working. What warrants denying them effective access to this public good by imposing large "user's fees" on them, as if this were just a highly desired private good to which they had no claim whatsoever?

Third, in a typical market, consumers have alternative ways of gaining access to a particular good or service. If I cannot afford a union plumber, then I can hire a nonunion plumber, or just pay my friend down the street who dabbles in plumbing. But this is not an option we have regarding health care. Health care professionals have used the power of government to award themselves monopolistic control over the delivery of such services. No one can just set themselves up as an amateur brain surgeon. There are excellent social welfare reasons why such restrictions are highly desirable, and consequently, virtually all of us approve such monopolistic control. But this does mean that the uninsured working poor have no alternative health care system to which they can turn. Of course, they can appeal to the charitable instincts of health professionals, but the response that has been elicited of late has been uncertain and inadequate. Further, a charitable request assumes that an individual has no claim whatsoever to a good, which would only be true if that good were wholly or primarily private.

Fourth, health care needs to be treated as a public good because it is the sort of interest that we all have, but that none of us as individuals are capable of looking after adequately on our own. Again, this is a purely contingent historical matter. Our colonial forebearers did not need the Food and Drug Administration to assure the purity and safety of the food and drugs they consumed. In simpler times, this was a matter that could be reasonably left to individuals. But in the highly specialized and complex society we inhabit, no one has the breadth of knowledge or time that would be needed to protect adequately their own interests in these matters. Hence, we treat this as a public interest and tax everyone accordingly. The same is true with regard to health care. No one can have all the medical knowledge or skill needed to treat oneself, nor can one even have the knowledge needed to assess adequately the abilities of those who would offer medical treatment to others. Hence, as a public interest, we regulate and restrict where and by whom health care may be provided. That interest could hardly be correctly described as being adequately protected as a public interest if 37 million citizens have no assured access to needed health care.

Fifth, we all have an interest in having access to adequate health care when we are older because there is a severalfold increase in the likelihood that we will need substantial amounts of health care then. But in the world of expensive high-tech health care, this is not an interest that most of us will be able to take care of by ourselves. Relatively few of the elderly can afford what would then be the very high costs of health insurance, if insurance companies were even willing to offer it to them at all. This is why the Medicare program was created. But do the nonelderly have any less of an interest in having secure access to needed health care? Not at all. However, the theory is that the nonelderly would be working, would have fewer health needs, and would receive health insurance as a workplace benefit. This fit the socio-economic reality of post World War II America for a couple decades. But that is not the reality we face today as America goes through the painful process of deindustrialization. Each year, millions lose their jobs and their health insurance through no fault of their own. Rather, this comes about because of massive economic changes that individuals themselves have little ability to control, just as the elderly have little ability to control the fact that they age and are exposed to more illness. So why should we not also think of health insurance for the nonelderly as a public interest?

To summarize, we have offered good reasons for saying that health care should be thought of as a public good and as an object of moral concern by society at large. But this still leaves us some distance from concluding that our present mechanisms for financing health care in America are so morally and economically flawed that they must be replaced by some system of national health insurance. We now offer preliminary evidence for that claim.

Just Health Care Financing: Critical Comments

Our present methods for financing health care in America are open to serious moral criticism. I offer the following considered moral judgments in support of that claim.

First, the existence of absolute administrative waste in the financing of health care in America is *prima facie* unjust. Any large and complex organization will have some amount of waste and inefficiency, which may simply have to be tolerated. We in the US, though, generate an enormous amount of administrative waste because we have a highly fragmented system for financing health care. Himmelstein and Woolhandler note that Canada, which has a single mechanism for financing health care, spends about 6% of all health care dollars for needed administrative services. But we in the US will spend about 16% of our health care dollars for administrative services.[23] That means that 10% of all our health dollars do not produce any health benefits for anyone. That means that in 1989, $61 billion were spent for wasteful and unnecessary administrative services—all those services associated with checking and rechecking whether a given patient was eligible for certain benefits under one or another of 2,000 health plans in the US. Further, all this checking has to occur in insurance offices, physician offices, employer offices, and several hospital departments. This is an enormous bureaucratic structure we have created in order to sustain "consumer choice" among these 2,000 health plans. And that is a cost to which no health benefits are attached. If there were no uninsured or underinsured individuals in America, and if there were no pressures to reduce escalating health care costs by reducing the "burden" of charity care, then this kind of waste might be morally tolerable. As things are, however, that loss of $61 billion in health care purchasing power very directly impedes access to the health care sytem for the uninsured. And that is unjust.

Second, haphazard and uncertain access to needed health care is *prima facie* unjust. As noted earlier, the uninsured are not entirely excluded from access to our health care system. They receive a substantial amount of charity care. But charity care is something that is freely given. Consequently, as competitive and cost containment pressures on hospitals mount, the certainty and adequacy of that charitable response are greatly compromised. If lack of access to that care merely resulted in inconvenience and a bit more discomfort and suffering that might have to be tolerated, then the failure of that charitable response would be morally tolerable. But when health care is needed, and when that need is not met in a timely fashion (as in the case of our woman from Detroit), and when the health consequences are devastating and irreversible, and when the moral consequences involve effective denial of equlity of opportunity, then a merely charitable response is totally inappropriate and inadequate from a moral point-of-view. As the President's Commission pointed out, there is a serious societal obligation that is not being met, and that we could hardly expect would be met so long as health care for the uninsured is thought of as a matter of charity.

The major source of uncertainty in providing "charity" care is our highly fragmented system for financing health care. Everyone sees charity care as someone else's moral responsibility, the result being the creation of these moral twilight zones in which the distinction between unfortunate and unjust health outcomes is mostly muddled. A good illustration of this is a recent story in the *American Medical News,* where it was pointed out that the poor suffer 10–15% more cancer deaths than the nonpoor because they delay getting appropriate medical care in the early stages of cancer when the likelihood of successful intervention is substantially greater.[24] The delay occurs because they would have to pay out of pocket for a physician visit and tests. Given our earlier analysis, this is unjust, though each of these deaths taken locally and separately will only appear to those around the deceased as being unfortunate. The fact that this first injustice is effectively hidden from public scrutiny provides reason to conclude there is a second injustice here as well.

We return to our main point: haphazard and uncertain access to health care is *prima facie* unjust. In our system for financing health care, this proposition is as true for the uninsured as for those insured through their place of work. A dynamic world economy causes turmoil in our job markets. Still, we have no reason for resisting those changes because both the national economy and the world economy are better off as a result. But we have no moral right to purchase this increased economic welfare at very great expense to the individuals who lose their jobs (or just change jobs), thereby losing their health insurance. In short, there are strong moral considerations for separating access to health care from having a specific job.

It is also noteworthy that insurance companies go bankrupt on occasion. This may be of small consequence for those in good health, who will be able to switch readily enough to another firm. But this will not be true for those who have serious health problems, such as the father of twin five-year old girls suffering from a rare blood disease. They each need a bone marrow transplant, but his insurance company has gone bankrupt, and a medical center in Kansas has said it will need $260,000 up front if the surgery is to be done.[25] It is clearly unfortunate that these girls are afflicted with this medical problem, but it is much harder to judge that it is merely unfortunate that they are being denied health care that could make the difference between life and death, and that this should depend on the vagaries of business judgment.

Third, the current tax system subsidizes middle class access to health care by failing to tax health benefits provided by employers, and this is unjust. In 1989, this subsidy (tax expenditure) amounted to $38–42 billion. This is money that state and federal governments could have collected if the value of health benefits was taxed as income. This represents a subsidy to the middle class because that insurance would cost 30–40% more if it were purchased with after-tax dollars.

There are two serious injustices in this arrangement. The first is that the working poor are denied this subsidy when they are denied health benefits by their employers. That means that they will have to purchase their health insurance with after-tax dollars, if they can afford it at all. In effect, this means that those who are already less well off will be made worse off, either by not being able to purchase health insurance or by having to pay much more for less adequate insurance (because they are denied the opportunity to purchase their insurance as a group).

The second injustice derives from the fact that the middle class has been given a powerful economic incentive by the government to purchase very comprehensive health benefits, which contributes greatly to the overall escalation of health care costs.[26]

Those escalating costs have precipitated all manner of defensive moves by both government and business to control escalating health costs for themselves. If these cost containment mechanisms simply represented efficiency-enhancing devices that applied to their own clients, there would be little that was morally troubling. However, these mechanisms often work to exclude from the health care system those who cannot pay for the care they need, e.g., only 40% of the poor in our society are Medicaid eligible because states establish stringent eligibility requirements to protect state budgets. That means that the working poor are funding with their tax dollars (at least in part) those mechanisms that will deny them access to a health care system their tax dollars helped to make possible. This is doubly unjust. What moral considerations would justify a system for financing health care that operated like this?

Fourth, patchwork reforms of the current health care system are costly, inefficient, not subject to effective cost containment mechanisms, and consequently, are unjust. At least fifteen states are trying to do something about the problem of the uninsured. The proposals are numerous because the uninsured are quite heterogeneous. They include state-mandated health benefits that would have to be provided by small employers, categorical additions of various needy groups to the Medicaid program, an income-graded Medicaid buy-in option for those above the poverty level, state-financed risk pools for those without insurance because of serious chronic illness, and so on. All of these proposals are well intentioned, but they are also politically unstable and subject to future erosion because they directly add to the cost of health care in the state, add to the complexity and cost of administering the system, and do nothing to control escalating health care costs. Further, these additional programs do virtually nothing to remedy the overall inequities in our health care system.

Fifth, there is no adequate moral justification for a multi-tiered health care system, at least as long as we are talking about medically necessary health care. If what I have argued for earlier in this section is correct, then having access to effective, medically necesary health care is a matter of justice. I will readily concede with Bayer and others that the concept of "medically necessary health care" is not a purely scientific standard. Rather, objective scientific criteria, medical judgment, and community consensus will all have to fill in the content of this standard in practice.[27] At the fringes of this standard, such a consensus may be difficult to achieve, but for our present purposes this will make no difference. The point is that whatever is adjudged to be medically necessary care ought to be distributed in accord with need and ability to benefit, not ability to pay, which is what actually occurs in a multitiered health system.

Consider the situation we now find ourselves in with regard to AZT. This is a drug that has been given to individuals with full-blown AIDS in the hope that this might prolong their life an additional year or so. Unfortunately, this drug costs $8,000 per year and has major adverse side effects in 50% of the patients given this drug. However, in late 1989, it was shown that the toxic side effects of this drug are reduced to 5% if it is given to patients much earlier in the course of the disease, before their immune systems have been severely compromised.[28] Apparently, this will also yield improved length of life. The practical problem, however, is that the number of people who would then need this drug would jump from 20,000 to somewhere in excess of 500,000. That, in turn, generates our moral problem. About 25% of individuals needing this drug will have the costs paid by their insurance. Another 25% may be eligible for state Medicaid support because they are impoverished (though this will vary from one state to another). As for the rest, their access to the drug will be utterly uncertain. Is there any kind of moral justification for this kind of variability in treatment when access to the drug will likely have the same beneficial effects for virtually all those 500,000 individuals?

There are numerous other inequities like this that are rife in our highly fragmented system for financing health care. The federal government, for example, has had in place a program that provided funding for AZT for those unable to pay, part of the argument being that this was an experimental drug. But continued funding for this program was in jeopardy even before the announcement of these new clinical recommendations, which would increase dramatically the total costs of the program to the federal government. But the equity issue is hardly resolved if the federal government does underwrite these costs, for the question can be justifiably raised as to why the federal government is not providing funding to other individuals suffering from other diseases who also could be helped to live longer by expensive drugs that they themselves cannot afford. This is the dilemma we face now when we provide public funding through the ESRD 1972 Medicare amendments for all patients in kidney failure who need either kidney dialysis or a kidney transplant. The federal government does not fund any other major organ transplant procedures. *Prima facie,* this appears unjustly discriminatory.[29]

A national health insurance program that is affordable may not be able to pay for all medically necessary care, given the pace of advance in medical technology and the cost of disseminating those technologies. That means that difficult and painful choices would have to be made, that some individuals will suffer unnecessarily or die prematurely if they had access to these technologies. Such results will be regrettable, but it is not obvious that they would be inequitable. For such difficult decisions would have been made in the light of the best cost benefit and cost effectiveness information available, not to mention a number of other morally relevant criteria, all of which would avoid discriminating among individuals on the basis of their ability to pay for this care. To the extent that that kind of impartiality can be achieved, everyone in our society would have been accorded equal care and respect.

Competing National Health Insurance Proposals: A Moral Assessment

In the remainder of this essay, I shall critically assess only two proposals for national health insurance: the Enthoven-Kronick proposal (hereafter EK) and the Himmelstein-Woolhandler proposal (hereafter HW).[30] Both represent fairly radical reforms of our system for financing health care, but neither exceeds the limits of the feasible. More interesting for our purposes is the fact that they represent competing conceptions of how health care justice ought to be embodied in a system for financing health care. The HW proposal reflects a more egalitarian conception of justice whereas the EK proposal is very much rooted in a libertarian conception of justice, though it is not at all a stingy libertarianism. This last point is important because their proposal would be lacking in intuitive moral appeal otherwise. They do see a necessary role for government in assuring effective liberty for all in gaining access to our health care system, though, as we shall see, this is well short of assuring equal access for all to our health care system.

Two Proposals: A Summary

EK describe the American health economy as a "paradox of excess and deprivation" in which we see ever increasing sums of money (both public and private) spent on health care while growing numbers of individuals who most need health care are denied it.[31] They view this system as being neither fair nor efficient. What they would recommend in its place is a system of near universal health insurance that would be provided through managed competition. They call attention to the research literature that suggests that HMOs (prepaid health plans) have been effective in reducing the cost of care from 10–40%. Their basic strategy is to build a system for financing health care that would have these plans at its core. They see this option as being increasingly acceptable to Americans because about 60 million Americans currently belong to managed care plans of one kind or another now. The theory is that these plans would have to compete with one another for clients, and they would have to do this by offering a better quality package of health benefits at a lower price than their competitors. In this way, current inefficiencies in the health system would be squeezed out, and escalating health costs would be brought under control. In order to limit the appeal of traditional insurance plans, tax deductions would only be offered to businesses for 80% of the costs of the average plan. Thus, those who wanted more expensive plans would have to pay for them with after-tax dollars, thereby denying public subsidies for markedly more inefficient forms of health insurance.

EK realize that few individuals would have sufficient time and background knowledge to assess the relative risks and benefits and efficiencies of these competing health plans. This task, therefore, would fall to "sponsors," usually larger employers who would have the capacity to critically assess these plans and to bargain for better prices for their employees. Still, the key to achieving their overall cost containment objectives is consumer choice. Sponsors would reduce the field of choice to a small number of competing plans, but then individual employees would have to make the ultimate choice for themselves in the light of their own risk/benefit or cost/benefit budgets.

Besides these private sector sponsors, each state would create a "public sponsor" agency that would cover small employers, the poor, and all those unable to obtain traditional insurance coverage. All employers would have to offer some range of health plans to all full-time employees and pay 80% of the cost of those plans. For part-time or temporary workers, employers would have to pay an 8% tax on those wages up to $22,500. Also, self-employed individuals would have to pay 8% of their adjusted gross income to the state. The federal government would pay the uncovered 20% of plan cost for those below the poverty level, while some gradated federal subsidy would be available to those from 100–

150% of the poverty level. Finally, there would be some modest deductible and copay requirements for individuals in order to assure the more efficient use of the health care system.[32]

Whereas EK see as a major virtue of their proposal its non-disruption of current patterns of health care financing, HW see this as inequitable and inefficient patchwork reform. What they propose instead is a single public plan that would cover all medically necessary services, which would include acute, rehabilitative, long-term, and home-care, as well as mental health services, prescription drugs and medical supplies, and preventive health services. Boards of experts and community representatives would judge what would count as medically necessary services. There would be no private health insurance at all. There would be no copayments or deductibles that would function as financial barriers to entering the health care system. As in Canada, hospitals would receive an annual lump sum "global budget," which would have to cover the operating expenses of the hospital for the entire year. Capital budgets would be separated from operating budgets in order to facilitate more rational and efficient regional health planning. Physicians would be paid on a fee-for-service basis, as in Canada, or they could be salaried through hospitals or HMOs. Monies now paid by employers in the form of health benefits would be captured in the form of a health tax on all employers. Medicare and Medicaid would cease to be since there would be only one health plan for all citizens.[33]

Some Assessment Criteria

I would recommend the following criteria for purposes of asessing our proposals: universal access, comprehensive coverage, elimination of administrative waste, effective cost containment, just rationing methods, equal treatment of all for similar medical conditions, maintenance of quality health care, fair payment to providers, equity and efficiency in financing health care, support for medical innovation, protection of professional integrity, protection of patient autonomy, overall fairness (space limitations require that only some of these criteria are discussed here, though the interested reader may write the author for the full discussion). I make no claim that this list is complete, though these criteria represent the central elements of health care justice with reference to health policy. In that respect, they represent a narrower perspective than what a generic health policy analyst would bring to the task of assessing health policy. To a large extent, these criteria represent the critical considered moral judgments that I offered in the prior section with reference to current policy in the US for financing health care. But I do not believe that these criteria have to be interpreted in a way that is intrinsically biased, say, toward a more egalitarian conception of health care justice. To assure this, I will weave into the discussion of most of these criteria the major objections to national health insurance raised by moral philosophers with a more libertarian bent. Specifically, I have in mind the work of Loren Lomasky,[34] Paul Menzel,[35] and Baruch Brody.[36]

Universality

The key question implied by this first criterion is this: is a specific proposal universal in the sense that the health needs of all are covered? The EK proposal falls short on this criterion because their proposal could do no more than "encourage nearly universal coverage."[37] They concede that migrant farm workers, the homeless, and others not attached to a place or a job would not be covered by their plan. These individuals would have to rely on a charitable response. And, as noted earlier, the uncertainty of this response is unjust when health care is needed.

More problematic would be the circumstances of high risk patients, individuals, for example, who are at high risk of cancer or heart disease (as suggested by family history), and who would likely incur large medical expenses. EK would "risk rate" such individuals and attach a higher premium payment to them so that they would not be discriminated against by competing health plans. That additional premium would not be paid by these individuals. Instead, it would be paid by the "sponsor." In many cases, that would be a larger employer who might then have an incentive for not employing that person in the first place. The extent to which this would be a serious problem and a serious injustice would depend on labor market conditions at a particular point in time. EK consider this problem but dismiss it for feasibility reasons.

More problematic still would be the circumstances of those who are seriously and chronically ill and unemployed, individuals who almost certainly will incur $50,000 worth of health expenses if they are able to access the health care system. The only sponsor to whom they would have access would be the public sponsor. I find it difficult to imagine what sort of "premium" would have to be attached to them to assure their access to a qualifying health plan. But I pass over that problem for the more philosophically significant issue, namely, that these individuals would only have access to some "basic" package of health services because they could not afford to pay the additional costs of accessing a premium health plan. That means that these individuals would only have access to a single plan inadequate to their health needs.

Note that EK have deliberately titled their proposal a "Consumer Choice Health Plan." Choice is a key moral virtue for them. Individuals are supposed to judge what they regard as costworthy health care and choose a health plan accordingly, just as with other consumer goods. But individuals have no choice in the situation I described. In fact, the economic reality would be the reverse of what we would normally expect. That is, the health plans would be choosing those consumers that seemed the healthiest and posed the least risk to the financial solvency of the plan. EK anticipated that this kind of market failure could occur, but they contend that the sponsors "could employ various tools and strategies to counteract the causes of market failure."[38] However, this really assumes a level of altruism on the part of

corporate sponsors, which might be there sporadically on behalf
of long-time employees, but which would hardly be elicited on
behalf of potential new hires to ordinary jobs.

If all corporate sponsors did their fair share in hiring individuals with higher than average health risks, then rough equity
would have been achieved in meeting and paying for those health
needs. But EK imagine that each state would be the public sponsor of last resort. They recognize that the state would likely have
more than its fair share of less healthy individuals, but they believe that the state would also be the sponsor for large numbers
of currently uninsured individuals who are quite healthy and
employed now by small businesses, the implication being that
this would permit greater spreading of risks and costs. However,
the scenario that would most likely evolve over a few short years
would be one in which more and more likely ill individuals would
be sponsored by the state, which would drive up costs and drive
out more healthy individuals who would get a better bargain in
the private sector. The state would then repeat the well known
policy moves associated with Medicaid. Higher costs would mean
higher taxes. But healthy taxpayers would be reluctant to fund a
"welfare" program for the ill. The alternative strategies are to
squeeze providers or to trim benefit packages. Either way, both
equitable access and quality of care are inappropriately compromised for those with greater than average health needs.

The economic logic in these matters is inexorable when you
have a competitive and pluralistic system for financing health
care. At least under more traditional forms of health insurance,
and in an environment where cost containment pressures were
moderate, less healthy individuals were more dispersed across
insurance plans. But in the EK model, the clear long-term tendency would be for significantly less healthy individuals to filter
down and cluster in the publicly sponsored plans. Recall that the
primary stated objective of the EK proposal is to achieve a better
balance of equity and efficiency in financing health care. We will

readily concede that this competitive model would achieve the expected efficiencies and cost savings, but the bulk of these savings would accrue to the private sector sponsors whereas disproportionate costs would accrue to the public sponsors. Unless substantial taxes were levied on these private sponsors to recapture those savings (which would be a strong disincentive to achieving those savings in the first place), equity of access to needed health care would be greatly compromised for those with above average health needs under public sponsors. Also, effective autonomy in the making of many health care decisions would be severely reduced for these individuals since most of the more preferred health plans would have become fortresses for the well.

This last point is worth one further comment. Lomasky, for one, starts with the premise that we are all roughly equal in our vulnerability to disease. So the insurance mechanism represents a rational response for spreading the risk of financial disaster and assuring access to adequate health care. However, he contends that this goal can be satisfactorily achieved through private insurance. There is no need for a national health insurance program with its liberty-limiting implications.[39] However, as we look to the near future we see dramatic increases in our capacity to make quite accurate long-range diagnoses of likely illness in individuals because of genetic susceptibility. For example, there is a single gene that is responsible for producing an enzyme that properly processes cholesterol in the body. Dysfunction in this gene means susceptibility to heart disease and a broad range of other vascular diseases.[40] The practical implication of such advanced diagnostic technologies is that much disease and illness would not be seen as random, unpredictable events that might be inflicted on any of us, which in the past made rational the spreading of such risk through an insurance mechanism. Instead, some individuals would be "doomed" from birth to some likely pattern of illness whereas others would be quite fortunate. The fortunate would have no reason to freely associate in health plans with the unfortunate.

This segregation of the most likely healthy from the most likely ill would greatly undermine the capacity of the likely ill to finance the health care they would need. There would be gradated clusterings of individuals with the least healthy and least wealthy at the bottom. In other words, access to health care would be determined primarily by ability to pay rather than some criterion related to health need. This is exactly the kind of inequity that EK had hoped to minimize by creating a system that would allow nearly universal access. The problem is that in a pluralistic system individuals will sort themselves out in ways that they see as being most economically advantageous to themselves, not in ways that preserve some level of desired or required equity. What was supposed to be the moral virtue behind pressing for universal access quickly becomes a moral sham under the EK proposal.

Comprehensiveness

The basic question behind this criterion is: does a specific proposal provide a comprehensive package of health benefits such that all needed health care is assured? This criterion has a deceiving simplicity about it that hides difficult issues in application, only some of which we can consider.

First, this criterion seems to imply that more care is better than less. That, however, is an empirical matter. There is considerable research evidence today that too many patients are the victims of too many tests and too many unnecessary surgical interventions.[41] Hysterectomies, endarterectomies, and coronary bypass surgery seem to be among the more frequently overutilized surgical procedures referred to—all of which expose patients to significant increased risks without corresponding medical benefits. Similarly, excessive testing can confuse medical judgment by increasing the likelihood of false positive test results, which in turn, can have adverse consequences for a patient. Too much health insurance can be dangerous to your health! This, however, does not count against the comprehensive type of national health insurance recommended by HW. For these problems can be remedied through carefully constructed quality assurance programs, and more careful research into the actual benefits of new medical technologies under a range of clinical circumstances.

Second, given continued rapid advances in the development and dissemination of new medical technologies, comprehensive health care for everyone might not be socially affordable. Health care could hijack our social priorities, especially if we were locked into a comprehensive national health insurance program. We might like to believe that the concept of health needs would provide the restrictions needed to prevent this hijacking. However, as Callahan has observed, our definition of health needs tends to be driven by whatever technologies are at the forefront of medical science. He writes, "No one thought, a century ago, that a person suffering from heart disease 'needed' a heart transplant; death was simply accepted."[42] We also seem to need bone marrow transplants, magnetic resonance imaging, and gene therapy. He also points out that, in our society, we operate with a broad, almost limitless definition of health, and a highly subjective notion of individual health need, due largely to the fact that ours is a liberal society.[43] This line of argument suggests that the EK proposal is preferable to HW because it would allow individuals to choose health plans that fit their individual priorities regarding health. That is, individuals would be free to spend "excessive" amounts of the money on health care, but in so doing, there would be no distortion of social priorities.

Recall that EK would repeal the tax advantages that currently stimulate the purchase of excessive health insurance. We might conclude that there is no ideally comprehensive health insurance plan, that this will always be relativized to individual judgments of what counts as "costworthy" health care.[44] However, the matter is not quite that simple. Recall that EK would not do away with the Medicaid program, though they would want to offer Medicaid recipients the option of buying into one of the plans offered by the public sponsor. Medicaid recipients would be given vouchers that would be equal to "the value" of the Medicaid package of services. Presumably that value would continue to vary from one state to another, as is now the case, which would continue to raise issues of broad societal equity. More importantly, this is a matter in which the recipients themselves would have no authoritative say. That is, they would not have the effective right to choose for themselves what counted as an adequate health care package for their needs. Instead, this is a choice that would be made by "others."

If there is a strong moral warrant for others in our society making such choices for themselves, then we have to ask what the competing moral considerations are that would justify our not recognizing a similar warrant in the case of health care choices for the poor. If we had good reason to believe that societal decision makers made health decisions for the poor that were strongly motivated by beneficent concerns, including the concern that the poor might not make sufficiently informed choices for themselves, then we might take exception to this paternalistic penumbra, but we could respect as morally legitimate the deeper moral motivation. However, if the current Medicaid program and its history are a good predictor of the likely future, then it is more probable that budgetary constraints rather than beneficent concerns will determine the adequacy or comprehensiveness of that benefit package. Or, to put it more bluntly, it will just not be a relevant consideration in state legislatures to ask whether the Medicaid benefit package is adequate to the health needs of the poor. If the Medicaid program is conceived of as a welfare program, as a matter of charity, then almost by definition anything that is done will be "enough," just as when I donate five dollars in support of some environmental cause.

To appreciate this last point we need to rectify an earlier point. We suggested that if individuals chose their own health plans with their own money, then this would avoid the skewing of social priorities that could happen if health care were entirely funded by public means. However, in reality there are substantial spillover effects from private sector health spending to the public sector. Specifically, if escalating health costs cannot be adequately controlled because of the continued broad diffusion of new and expensive health technologies, then access to these technologies has to be considered for those covered by public programs. But this would result in a number of unattractive choices in the public sector. Either taxes would have to be raised to cover these increased health costs for increased health benefits for the poor, or the budget of other public programs would have to be trimmed to divert more public resources into health care. The two other largest items in typical state budgets are public education (higher education, especially) and highways, both very important public goods to the middle class, who see themselves as prime beneficiaries of these goods, whereas they would not see themselves as beneficiaries of welfare programs. Given these choices, the most likely loser would be equitable access to adequte health care, especially for the poor and near poor.

Here we need to recall another of our earlier points, namely, that efficiency, in the sense of effective health care cost containment, may not be achievable without a solid commitment to equity. Both values may be adequately protected only if they are traded off against one another within the confines of a single system for financing health care. As things are now, the middle class achieves, in part, its health care cost containment objectives by trading off access to health care for the poor. That is, the less they pay in taxes to support Medicaid, the more they may spend on their own health priorities, which will help to fuel escalating health costs for all. But if all are committed to a single package of health benefits, then there will be the social motivation needed to identify and purchase only health care that is judged to be costworthy in broad social terms. Conceived in this way, the political process may prove to be more trustworthy, more rational, and more just than we might initially imagine. In fact, we would probably find here the closest approximation in reality to the Rawlsian veil of ignorance.

We might fear that all the specialized "disease interests" would undermine or grossly distort the political process. But the political reality is that the vast majority of people in any given year are quite healthy and quite ignorant of their most critical future health needs. Under such circumstances, I would imagine that a quite comprehensive social health contract would emerge from any bargaining process since it is the same contract that we would all have to live under. But health care would not be able to hijack the public budget, for the broad middle class would want to protect those interests as well. Some difficult trade-offs would be inescapable, but we would agree on a broad range of possible interventions in specific circumstances that would be excluded from the health contract because they were not costworthy—the marginal benefits that were just not worth the social costs as determined by our healthy rational selves at an earlier point in time. More importantly, there would be a keen realization that future health improvements would have to be purchased through increasingly efficient future health technologies; and so there would be strong political and economic pressure for supporting the dissemination of those new health technologies that met strict social efficiency criteria.

There is yet another objection that can be raised in connection with our comprehensiveness criterion. Lomasky points out that if there is a single publicly funded program for health insurance, then some people will be forced to pay taxes to provide health procedures against which they have strong religious objections. Lomasky mentions Medicaid funded abortions,[45] but there are dozens of other examples from the field of reproductive technology or gene therapy or terminal care. He believes that "divisiveness can be minimized by leaving decisions in the hands of private citizens."[46] However, this objection really misses the mark if he thinks this would be accepted as an adequate response by members of the right-to-life lobby, for their objective is to ban these medical practices altogether from our society, whether those practices are publicly or privately funded. So this does not really count against a national health insurance program.

A more difficult objection would involve the denial of funding for certain expensive life-prolonging procedures in the case of terminal illness under a national health insurance program. Such rationing decisions might be necessary in order to achieve certain cost containment goals. Thus, we can easily imagine that patients in a persistent vegetative state would be denied either respirator support or artificially provided nutrition and hydration after it was confidently judged that they would never recover from that state. Or we can imagine a seventy-year-old individual in the end stages of pancreatic cancer, desperately wanting to live, who wanted access to a therapy that cost $100,000 and offered a 25% chance of an extra six months of life. It is hard to imagine that such therapeutic options would be funded under a national health program.

But how could a liberal society deny individuals the right to make such choices for themselves? Lomasky contends that these ought to be matters of personal decision and not transformed into public policy questions. As he observes, "Regulation is, in essence, inflexible."[47] If there are budgetary limits that must be respected, then there will have to be formulas that govern the use of expensive life-prolonging resources near the end of life. But those formulas will have to restrict discretionary authority in order to avoid being subverted from within. Further, these formulas would be worked out by a "distant bureaucracy" ignorant of the medical and emotional detail of the life at stake. Lomasky sees this as a subversion of the integrity of the practice of medicine itself. He writes, "Health services are not provided to faceless pathological syndromes but to persons whose preferences and circumstances are endlessly varied and complexly interrelated. If responsiveness to individual circumstances is something worth prizing, it is morally obtuse to restrict attention only to data that can be quantified and processed by technocrats."[48] Lomasky concludes this objection by recognizing that efficiency requires consistency in the application of rationing formulae. However, he then questions whether such efficiency is a virtue to be consistently prized over "compassionate willingness to accede to individual needs."[49]

In fairness to Lomasky, we must note that the health care system has changed dramatically in ten years. Lomasky wrote when retrospective payment mechanisms were dominant in health care. That is, physicians provided any health services to patients that they judged would improve their health condition, and they were paid for each of these services. The more physicians did, the more they earned. This has obvious inflationary potential, especially when medical science makes available to physicians an ever expanding array of beneficial health technologies. Hence, the consensus of health economists today is that prospective financing mechanisms, fixed budgets rather than open-ended financing, are necessary to discipline medical judgment and thereby contain health care costs. This is reflected in the fact that the health plans envisioned by EK are all HMO-type entities with fixed annual budgets and fixed annual contributions by plan participants. The practical upshot of this is that if we do not have a single national health insurance program, then it will be distant private sector bureaucrats armed with equally inflexible rules and regulations who will coldheartedly deny our end-stage pancreatic cancer patient the intervention that represents a possible six extra months of life. In other words, if we are operating with a prospective payment mechanism, then neither physicians nor patients can be granted broad discretionary authority in the kinds of medical circumstances we are considering, at least if we do not wish to defeat the whole point of the prospective payment mechanism.

The public/private sector distinction is simply not relevant. We can imagine that there would be plans in the private sector that would be more or less generous in their benefit packages. But this point too misses the mark, for even in the most generous packages there will still be limits. The reader is asked to consider whether he or she would want to be part of a health plan in which all participants are committed to doing everything medically possible to prolong their lives. That is, each member would have unlimited access to the aggregated financial resources of the plan. That would be the practical equivalent of giving each member of that plan unlimited access to one's private bank account, for ultimately, that is the source of those aggregated resources.

Of more serious concern is Lomasky's reference to distant bureaucrats making rationing decisions. There ought to be legitimate moral concern when some impose rationing decisions on others who have not freely consented to such limitations. This is the situation Lomasky is asking us to imagine under some system of national health insurance. However, why do we have to imagine it this way? Why can we not imagine that individuals would freely impose these limitations on themselves through a democratic process of discussion and decision making? What would emerge from such a process would be a mechanism of mutual restraint mutually imposed for mutually agreed on objectives. There would continue to be both bureaucrats and physicians who would continue to serve as gatekeepers of the system, but they would simply be the practical organizational mechanism needed to implement the rationing arrangements that participants will have earlier agreed to impose on themselves.

The basic insight needed to trigger this conversational process is recognition that the aggregated pool of health insurance funds is not a gift from some generous benefactor (which is the perception encouraged by health benefit packages tied to employment). Rather, that fund is derived from money that would otherwise be mine to spend as I wish. The second needed insight is that the insurance mechanism represents the most rational response to assuring access to costly needed health care. But that does set up the "tragedy of the commons" problem. If everyone has absolute freedom to use up the resources of the commons, then the commons will be quickly depleted (as well as my own private resources in this case). The third needed insight is that we will need some system of rules for the prudent use of the commons, and these rules will bind equally all those who have contributed to creating the commons. That is, no one has intrinsically a stronger right to life than anyone else. We might call this the "reciprocity condition." Whatever rights we wish to grant ourselves by way of accessing the medical commons will have to be granted to all other participants. Thus, if we are unwilling to allow the stranger in our group to use $100,000 for a 25% chance at six extra months of life, then we will have to be willing to deny ourselves and those we care about that same treatment under similar circumstances.

In order for a system like this to work, there would probably have to be several thousand clinical protocols that would have to be articulated. It would require considerable research and technical expertise to generate these protocols. And it is impossible to imagine that these would emerge from some sort of democratic process of bargaining and negotiation. Instead, the democratic conversation I envision would seek to achieve agreement on a more limited number of cost/benefit, risk/benefit, and other value trade-off considerations that would then govern the selection of clinical protocols governing rationing in clinical practice. The reader may wonder at this point how we might achieve a suitably impartial perspective for purposes of making these choices. But, as I noted earlier, the vast majority of us are situated behind the practical equivalent of a Rawlsian veil of ignorance when it comes to our future health needs. The practical implication of this is that we would want to protect our access to a very broad range of health services, so it would be very unlikely that we would discriminate against individuals afflicted with specific diseases. But because there was only a certain fraction of GNP that we would be willing to commit to health care (which would also be the same fraction of our personal income), we would have to make some difficult trade-offs with respect to our future access to health care. Although we would likely continue to invest in medical research, it will no longer be a sufficient criterion for the dissemination of that research that it yield technologies that do some medical good. Rather, the good will have to be substantial and costworthy from the perspective of our social budget.

We can now return to Lomasky's concern about compassion being squeezed out of our health care system by the impersonal demands of efficiency. This objection is really a moral red herring. It distracts us from acknowledging the requirements of justice under these circumstances. Moreover, it is misleading in its implication that compassion is best embodied in life-sustaining technologies. On this latter point, it is eminently arguable that hospice care represents a more compassionate response to the needs of the terminally ill than a mega-dose of some advanced chemotherapeutic agent that offers a 25% chance of six more months of life. On the former point, it must be noted that the moral luster of compassion purchased at the price of justice is greatly tarnished.

Unless we believe that new resources are magically and costlessly poured into the health care system, then the compassion displayed by a physician who violates one of the rationing protocols will come at the expense of other patients who will be denied needed health care to which they will have a just claim. Those patients may not be identifiable now, and it may be the case that they will not be precisely identifiable in the future. That makes the exercise of such compassion psychologically easier but no less unjust, for the losses will still be real. These invisible patients have just claims because they contributed to the system with the understanding that these rationing protocols would be adhered to by all who were part of the system and who had benefited from the fact that others in the system had earlier adhered to the protocols. They would have benefited through the savings that would have been achieved. Consequently, they have no right to expect or to accept compassion from caregivers that involves the allocation of health resources unjustly. In the final analysis, if compassion like this could be exercised in this system, it would undermine the system itself, most especially the broad social and individual benefits such a system would make possible.

Finally, we need to ask whether the Rawlsian-like health contractor I have imagined would rationally prefer the EK pluralistic system for financing health care to the monistic HW proposal. At least one strong rational consideration would tilt a decision toward the latter proposal. This is the risk insurors refer to as "moral hazard," which would threaten the long-term stability of the EK proposal. Specifically, individuals who knew themselves to be in good health for the coming year would purchase less expensive health plans for that year, but then switch to a plan with a more generous benefit package when they anticipated a serious medical problem. This is unfair to all those who would be long-term members of that more generous plan. This kind of gaming is an inevitable and costly element of any pluralistic system. The opportunities for that are eliminated in a monistic system. Further, no claim can be justifiably advanced that we are denying someone an important moral right when we deny them the opportunity to game the system.

Still, some libertarians might argue that individuals who are more well off ought to have the right to spend their money as they wish, that if they wish to spend their money on health care that is not costworthy, then no one should deny them that right. This is a reasonable argument, and consequently, I would be inclined to allow individuals to spend their own money outside the national health plan so long as such expenditures did not compromise the health welfare of those within the plan. But one of my working assumptions is that the national health insurance plan would be very comprehensive in covering costworthy health needs, so that it would really be noncostworthy health needs/wants that these individuals were seeking to satisfy. It is theoretically possible that a small private insurance market would reemerge under these circumstances. But this might be only a theoretic possibility, for it is difficult to imagine an affordable insurance plan designed for those who want virtually unlimited access to noncostworthy health care.

To conclude this discussion of the comprehensiveness criterion, there are a number of unanswered questions that remain. For example, to what extent should the voluntarily generated health needs of individuals be included or excluded from a national health insurance program? We have in mind here needs that would be associated with alcoholism or smoking or drug abuse or mountain climbing or eating junk food and so on. There are a lot of complicated philosophic and political and scientific and conceptual issues that would have to be addressed here.[50] Second, under national health insurance, who is it who should have access to experimental medicine, promising interventions that are at the edge of evolving medical technologies? And how exactly do we go about drawing the line between truly experimental medical therapies and medical therapies that have proven "effective enough" that they ought to be made generally available to the relevant patient population (and covered by our national insurance policy)?[51] Most insurance companies today still classify major organ transplants and bone marrow transplants as experimental medicine (for reimbursement purposes), though these technologies promise at least five years of survival to a majority of patients who have access to them. Third, how expansive should the coverage be for "mental illness" in a national health insurance program when there is so much disagreement among mental health experts regarding what should be described as mental illness, not to mention what would count as appropriate and effective therapy for whatever is identified as mental illness?[52] Fourth, how expansive should a benefit package be for long-term care and home care services under a national health program when many of these services are now provided "for free" by friends and relatives of elderly individuals? These are all difficult and complex questions, but they are equally so for both the EK and HW proposals. In that respect these questions do not provide discriminating considerations.

Just Rationing Methods

Our key question here is: to what extent does a specific proposal implement rationing decisions in ways that are fundamentally fair, i.e., patients are protected from rationing decisions that are arbitrary or capricious? Further, to what extent are these rationing mechanisms visible rather than invisible? And are they a product of a broad public consensus? My primary philosophic claim is that invisible rationing mechanisms, by which I mean rationing devices that hide from effective public scrutiny either the fact that rationing has occurred or the basis on which it has occurred, are *prima facie* unjust. They are unjust because they violate one of the core elements of our shared conception of justice, what Rawls refers to as "the publicity condition."[53] The basic idea is that when we have just public policies, there is no need or justification for hiding the allocative processes. I have argued for this claim elsewhere.[54] Most of the rationing that occurs in health systems today is effected invisibly. This is as true in Britain as in the US as in Canada.[55] Calabresi and Bobbitt have done admirable work in explicating the political and psychological circumstance that make invisible rationing an attractive option.[56] Unfortunately, they also argue that it is a morally defensible option. The HW proposal rejects in principle the idea of invisible rationing. EK do not speak directly to the issue; but to the extent that they trumpet as a virtue of their proposal the fact that it disrupts so little of our present health financing arrangements, we may reasonably conclude they would accept at a minimum the moral legitimacy of that form of invisible rationing that is represented by an individual's ability to pay.

Lomasky and Brody, both libertarians, are critics of national health insurance because it would make rationing decisions both inescapable and visible objects of explicit public choice.

We start by considering several objections raised by Brody. He rejects what he calls "the emerging standard view."[57] Its chief elements are that we need to contain health care costs by eliminating waste through regulation and competition and more emphasis on preventive health care. But this will not be enough, and so we will have to engage in rationing, which means that some individuals will be denied health care that they both want and from which they would likely benefit.[58] Brody will accept the fact that there are limits to what can and ought to be spent on health care, but these are limits to be established by individuals in the light of their own command of resources. For Brody, this is not rationing, for that term implies that someone else, i.e., a government bureaucrat, is restricting an individual's access to health care. So his first objection to rationing in a national health insurance program is that it would violate individual autonomy. He would prefer to give the poor either cash or vouchers to make their own trade-offs.

His second objection is that rationing necessarily involves deceit, either explicit or implicit. He contends that it is not sufficient that rationing policies are discussed publicly. His concern is what happens in the doctor/patient relationship. He contends physicians will be strongly motivated to minimize awareness on the part of both patients and their families that rationing is occurring.[59]

Brody's third objection is that there is no rational basis for making any rationing decisions, whether as a fraction of GNP or for ordering our allocation priorities within an overall health budget. He makes specific reference to a screening program for pregnant women aimed at preventing neonatal herpes in their infants at a cost of $1.8 million per detected case.[60] How do we know whether this is an investment we should or should not make? Here, Brody echoes a theme from Engelhardt, namely, that there are limits to reason.[61] The conclusion Brody draws is that this provides sound moral ground for letting individuals make their own choices, thereby respecting their values and priorities, rather than just having some set of social priorities imposed on them.

Brody's final objection is that rationing, in many instances, will involve a diminishing of the value of human life, especially if a public policy publicly proclaims that there are some lives that are not worth saving.[62]

Going back to Brody's autonomy objection, our response is that the autonomy offered the poor through health vouchers or cash is both illusory and inadequate, and hence, unjust. Perhaps the most critical question pertains to who sets the value of the voucher, since we have to assume that there will be large numbers of poor individuals with above average health needs for whom a voucher of only average value will be seriously inadequate. Brody says nothing to suggest that the poor themselves would have the right to set that value. On the contrary, it would be a government bureaucrat carrying out the will of the legislature that would likely do this. Now why should we not think of this as rationing in the morally objectionable sense that Brody has in mind? For the predictable result will be that the poor will be denied needed beneficial health care. Further, it certainly appears that this represents a devaluing of the lives of the poor since there will be tens of thousands of life-years lost each year due to denial of care for lack of ability to pay. These are lives that could have been saved had society made the choice to fund those vouchers more generously. What Brody would have us believe is that these are really unfortunate results of the impersonal workings of the market in conjunction with the impersonal workings of a nature that has inflicted some disease on the poor. But this is surely a morally disingenuous description of the working of markets. Markets may not be subject to the personal control of individuals, but they are surely capable of being subjected to larger social controls. Markets here are really instruments of social rationing.

On the assumption that there are limits to what we can and ought to spend on health care as a society, and on the assumption that health needs will continue to expand primarily because of continued rapid advances in health technology, then rationing is inescapable. And rationing will necessarily mean limits on individual autonomy. But in my judgment, what gives moral legitimacy to such limits is that they represent fair or just restrictions on the autonomy of all that all have imposed on themelves. If Brody is really concerned about protecting the autonomy of the poor in matters of access to health care, then the choice he would offer them is between vouchers under a pluralistic EK plan or participation in the universal health plan of HW. It is very difficult to imagine that well informed poor individuals would choose anything other than the universal plan. Apart from improved benefit levels, the most compelling moral consideration would be that the same rationing protocols would apply to all.

We recognize that those who are currently advantaged by virtue of having superior access to health care would not acquiesce as readily to such a universal plan since they would be exposed to the risks represented by rationing protocols, and these are risks from which they are now able to insulate themselves. But we also need to recognize that this is a biased perspective, and that they would see their choice situation quite differently if they were placed behind a veil of ignorance that assured impartiality. It also needs to be mentioned that the autonomy of the currently well off would not be unjustly compromised by the HW plan because under both our current health insurance system as well as the EK plan the more well off are really buying noncostworthy health care at social expense. That is, the less well off are really worse off in terms of access to health care because there are few social controls on the purchase of noncostworthy health care, even though much of this care is purchased at social expense.[63]

Brody's second objection to rationing is that it will involve deceit and dishonesty in practice, especially in the doctor/patient relationship. For the sake of argument, let us assume this is what would likely happen. Is this something that should count against the HW proposal because physicians would be gatekeepers charged with implementing clinical rationing protocols? Clearly, it should not because physicians would have to be gatekeepers under the EK proposal as well since there would always be fixed prospective budgets in those health plans. Moreover, there would surely be more powerful incentives for invisible and potentially dicriminatory forms of rationing under the EK proposal because all these health plans would be competing with one another for clients. It is highly improbable that any of these plans would push as a selling point the fact that they were especially skilled in making rationing decisions. More likely would be a large newspaper ad I noticed headed "$94,000 Smile." The ad shows a young mother and smiling baby and explains that this baby needed 15 weeks of hospitalization and nine specialists as consultants. Still, "Michigan HMO Plans *paid every penny.*" [64] The intent of the ad is blatant. No one would read that and think they were entering an arrangement in which any rationing decisions at all would be made.

Brody minimizes the value of the explicit rationing protocols that would be part of the HW plan because there is no guarantee that such protocols would be brought to the attention of patients in a specific clinical encounter. But at least those protocols would exist and would be generally publicized. In the case of the EK proposal there is no guarantee that such protocols would exist formally, much less that they would be honestly presented to plan participants at the time they considered joining the plan. This would not only vitiate informed consent at the most fundamental level of health decision-making (thereby violating consumers' autonomy rights), but it would also allow for the making of arbitrary and capricious rationing decisions. Moreover, even if there were formal protocols in some plans, there would be no guarantee of consistency from one plan to another. The moral and political virtue of the HW plan is that everyone could be reasonably confident of fair treatment because the same rationing protocols governed everyone's medical care.

Brody's third objection is that there really is no reasonable basis for these rationing protocols; and consequently, we ought to just let individuals decide for themselves what they regard as costworthy health care in the light of their own weighing of risks and benefits. But this objection is overstated. I concede that there is no uniquely rational ordering of health priorities to which a just and decent society ought to commit itself. Nor is there a uniquely rational ordering of health priorities at the individual level at that point in time when an individual must purchase health insurance. At both the individual and societal levels, there is considerable ignorance and uncertainty. That is what motivates and justifies the purchase of health insurance in the first place.

But our real concern, morally speaking, is that rationing decisions are made fairly, in a manner that is wholly impartial. This requirement can be met if we employ the best scientific knowledge we have at a given time, the best in the way of generalized clinical judgments, and open processes of social inquiry to assure that all pertinent value considerations are taken into account. If we can put into place such fair and impartial methods of inquiry, then the results would be fair enough and reasonable enough, even though they would not be uniquely rational.

Still, the libertarian may assert that justice is not the most important moral consideration here; rather, it is the preservation of individual autonomy, just as we have come to recognize at the clinical level. Imposing some universal health plan on all with a single set of rationing protocols is at best paternalistic, at worst unjust. However, the comparison with patient autonomy in the clinic is inapt. Patients there have a clear right to make risk/benefit trade-offs in the light of their own life plans and value hierarchies, at least to the extent that it is only their own welfare that is at stake. But when they must access a limited pool of social resources in order to effect their choices, which will be true under either the EK or HW plan, then the agreed on rights of others will legitimately restrict any one individual's access to that pool. In effect, the judgment is being made that that individual's current health needs are of lower priority than other health needs that must be met through that pool of resources, and that that individual had earlier agreed to that ordering of priorities.

So far as clinical decisions are concerned, an individual has a clear right to change his or her mind about his or her preferences at a critical moment, e.g., deciding to see what life on a respirator is like. But no such right can be claimed with regard to allocation priorities, especially when such a change of mind reflected a partial and advantaged perspective that would result in making unjust claims on a common pool of resources. Further, what we have to imagine is that a rationing scheme has been in place for some time, that current individuals have benefited from the savings achieved because prior individuals gave up potentially beneficial but noncostworthy health care in the past. What would then justify current individuals not keeping their end of the bargain?

Brody's fourth objection is that rationing reflects a morally obectionable devaluing of human life. This objection does have some moral bite, especially if we interpret this in a utilitarian spirit, and especially if we have in mind treating the old and terminally ill as if they were no more than worn out machinery to be discarded. But there is no logically compelling reason why rationing must mean this. Both Norman Daniels[65] and Daniel Callahan[66] have articulated age-based rationing schemes that are fundamentally respectful of the autonomy of each of us in these matters, because we make these rationing choices for ourselves. Space does not permit describing their views at length, but the general idea from Daniels is that we have to imagine ourselves allocating some limited bundle of health resources over the course of our whole life (as opposed to imagining the young making rationing decisions for the old). What we would probably prudently choose is to maximize our access to lifesaving medical resources at earlier stages in our life in order to improve the likelihood of our reaching old age, at which point we would give up our access to some range of such resources in exchange for better home care and long-term care, which are the needs that would likely be greatest at that point in our lives. Note that the success of this scheme requires a stable commitment to the scheme over the course of a life. This is an objective that would be more readily met by the HW proposal than EK, the reason being that the latter proposal would allow costly gaming by individuals as they jumped from one health plan to another in the light of their personal knowledge of their likely future health needs. It does not seem that a society treats anyone unjustly if it prevents that kind of gaming.

Overall Fairness: Some Concluding Remarks

To what extent are our current proposals a substantial correction of the obvious and widely agreed on injustices in our present system for financing and delivering health care services? The HW proposal does seem to correct all of those injustices identified in the first part of this essay whereas the EK proposal leaves most of them uncorrected. Specifically, the EK proposal does nothing to eliminate administrative waste in the system (and may even add to it in the long run), retains the connection between employment and access to health care with the instability that implies for access to health care, and retains a multitiered system with all the inequities and uncertainties in access to care that that implies. Further, the EK plan is only a "nearly universal" plan because it offers incentives for the uninsured to avail themselves of health insurance, but there is no guarantee that all will join. That means there would be some level of charity care to be provided with all the problems that entails in determining who is responsible for bearing that burden. Finally, there is little empirical evidence to suggest that the EK proposal would achieve the cost containment/efficiency objectives for the health care system as a whole, which is what was supposed to be one of its major virtues.

Here we need to recall that this entire essay is being written from the perspective of nonideal justice, from which feasibility must be a serious practical consideration. Thus, in their defense, EK write,

> To be politically viable, a proposal for universal health insurance must respect American cultural preferences for pluralism, diversity, local solutions, and individual responsibility. It must consider the preferences of providers and consumers for a variety of systems and styles of care. It must not provoke the strong opposition of large or important groups."[67]

They add that they see their model as being more "adaptable" than a public sector monopoly, by which they mean it would be less bureaucratic and more responsive to patient needs and preferences. Again, this overall response has something of a moral coloring to it since there is at least implicit a strong commitment to respect for personal autonomy. However, from a more critical perspective, we need to ask whose autonomy is being protected. Clearly, it is the autonomy of the middle and upper classes. Individuals from lower socio-economic strata who belonged to any of the publicly sponsored plans would have a lot less in the way of effective autonomy in accessing health care. That diminishes the moral luster of any appeal to autonomy. In addition, the superior autonomy of the middle class is being purchased at the expense of equitable access to health care for those who are poor or who have above average health needs. Finally, it seems reasonable to ask whether we would ever have had civil rights legislation or strong environmental legislation if a requirement for such legislation was "no strong opposition from large or important groups." We need to emphasize that nonideal justice is not a warrant for moral timidity and complacency.

In conclusion, we can only make comparative judgments in this nonideal framework. Our objective was to determine which of our proposals achieved a better regional reflective equilibrium in the problem/decision sphere we defined, i.e., just health care financing in America. Specifically, which proposal achieved a better balance of four major value considerations (cost containment/efficiency, equity of access, preservation of high quality, and autonomy of consumers and providers in health care), especially the numerous considered moral judgments subsumed under each of these values? Contrary to Brody's claim that greater emphasis on autonomy considerations will achieve the requisite balance and resolve the cost containment/equity problems that precipitated the discussion of national health insurance, I have argued that greater emphasis on equity considerations will achieve that balance, which is best represented in concrete terms in the HW proposal.

Notes and References

[1]Rashi Fein. (1989) *Medical Care, Medical Costs: The Search for a Health Insurance Policy* (Harvard University Press, Cambridge, MA); Samuel Levy and James Hill. (1989) "National Health Insurance—The Triumph of Equivocation," *The New Enaland Journal of Medicine,* **321**, 1750–1754; Stephen Shortell and Walter McNerney. (1990) "Criteria and Guidelines for Reforming the U.S. Health Care Sytem," *The New Enqland Journal of Medicine,* **322**, 463–467; David Kinzer. (1990) "Universal Entitlement to Health Care: Can We Get From Here to There?" *The New England Journal of Medicine,* **322**, 467–470; Roger Battistella. (1989) "National Health Insurance Reconsidered: Dilemmas and Opportunities," *Hospital and Health Services Administration,* **34**, 139–156.

[2]Testimony given by Robert Bitterman, MD to The Governor's Task Force on Access to Health Care in Michigan (July 27, 1989) at Gaylord Michigan Meeting, pp. 2–3. I served as the staff ethicist for this task force. The complete testimony is contained in the final report of the task force.

[3]The term "reflective equilibrium" belongs to John Rawls, for whom the term does not seem to be much more than a promising seminal notion. *See* his *A Theory of Justice* (Harvard University Press, Cambridge, MA, 1971), 20–22, 48–51, 579. Norman Daniels has probably done more than anyone else to flesh out this notion, but much still remains to be done. *See* his "Wide Reflective Equilibrium and Theory Acceptance in Ethics," *Journal of Philosohphy*, **76**, 256–282 and "Reflective Equilibrium and Archimedean Points," *Canadian Journal of Philosophy*, **10**, 83–104. The intellectual affinities are with coherentist theories of truth and justification in ethics. The concept I wish to introduce is that of a "regional reflective equilibrium," a balancing of competing considered moral judgments and relevant theoretical considerations within a specified field of inquiry or problem/decision sphere. Our field is "just health care financing." And the global value considerations in need of balancing are equitable access to health care, high quality health care, effective cost containment/ efficient resource utilization, and respect for patient and professional autonomy.

[4]The policy options I have in mind include Senator Edward Kennedy's *Basic Health Benefits for all Americans Act*, which would mandate that all employers provide a basic health benefit package to their employees; the Enthoven-Kronick proposal outlined in their two-part article "A Consumer-Choice Health Plan for the 1990s: Universal Health Insurance in a System Designed to Promote Quality and Economy," *The New England Journal of Medicine*, **320**, 29–37, 94–101; the proposal from the American Medical Association under the title *Health Access America;* the proposal from the National Leadership Commission on Health Care published by the Univerity of Michigan Press under the title *For the Health of a Nation* (1989); the proposal from the Heritage Foundation under the title *National Health Care System for America* (1989), edited by Edmund Haislmaier and others.

[5]*See* David Himmelstein and Steffie Woolhandler , et al. (1989) "A National Health Program for the United States: A Physician's Proposal," *The New England Journal of Medicine*, **320**, 102–108.

[6]Robert Evans. (1986) "Finding the Levers, Finding the Courage: Lessons from Cost Containment in North America," *Journal of Health Politics, Policy and Law*, **11**, 585–615, at 588–589.

[7]Robert Veatch. (1986) *The Foundations of Justice: Why the Retarded and the Rest of Us Have Claims to Eauality* (Oxford University Press, Oxford, UK).

[8]H. Tristram Engelhardt. (1986) *The Foundations of Bioethics* (Oxford University Press, Osford, UK), especially chapter eight.

[9]Norman Daniels. (1985) *Just Health Care* (Cambridge University Press, Cambridge, UK).

[10]*See* James Todd, President of the AMA. (1989) "It is Time for Universal Access, Not Universal Insurance," *The New England Journal of Medicine*, **321**, 46–47.

[11]*See* Shortell and NcNerney, *op. cit.*, 463–467.

[12]John Rawls, *op. cit.*, 279.

[13]*Ibid.*, 279.

[14]For full details describing the framework of nonideal justice, the reader may *see* my earlier article (1987) "DRGs: Justice and the Invisible Rationing of Health Care Resources," *Journal of Medicine and Philosophy*, **12**, 165–176.

[15]For a more complete description of the problem of those without health insurance, *see* Donald Cohodes. (1986) "America: The Home of the Free, the Land of the Uninsured," *Inquiry*, **23**, 227–35; or Gail Wilensky. (1987) "Viable Strategies for Dealing with the Uninsured," *Health Affairs*, **6**, 33–46.

[16]I first introduced the notion of a moral twilight zone in my paper "Moral Implications of the Shifting Power Equilibrium in Hospitals: From Medical Monopolization to Managerial Mandaranism," presented at the annual meeting of the Society for Health and Human Values in Chicago (November, 1988), later presented at the national meeting of the American Hospital Association in Chicago (August, 1989). For a summary of my paper *see* "Power Shifts Lead to Moral Twilight Zone for Hospitals," in *Hospital Ethics*, **5**, 1–4.

[17]Norman Daniels, *Just Health Care, op. cit.*, 226.

[18]Paul Starr. (1982) *The Social Transformation of American Medicine* (Basic Books, New York, NY), chapter three.

[19]Norman Daniels, *Just Health Care, op. cit.*, Chapters 2 and 3.

[20]The President's Commission for the Study of Ethical Problems in Medicinc and Biomedical and Behavioral Research. (1983) *Securing Access to Health Care*, Vol. I (Government Printing Office, Washington, DC), 17.

[21]*Ibid.*, 4.

[22]Leonard M. Fleck. (1989) "Just Health Care (I): Is Beneficence Enough" *Theoretical Medicine,* **10,** 167–182.

[23]David Himmelstein and Steffie Woolhandler. (1986) "Cost Without Benefit: Administrative Waste in U.S. Health Care," *The New England Journal of Medicine,* **314,** 441–445.

[24]Deborah Pinkney (1989), "ACS Report: Poor Caught in Cancer Trap," *American Medical News* (July 28, 1989), 1, 38–39.

[25]*The Chicago Tribune* (August 6, 1989), 7.

[26]Martin Feldstein (1983), *Health Care Economics,* 2nd Ed. (John Wiley and Sons, New York, NY), 480–557.

[27]Ronald Bayer, Daniel Callahan, et al. (1988) "Toward Justice in Health Care," *American Journal of Public Health,* **78,** 583–588, at 583.

[28]Paul Volberding, et al. (1990) "Zidovudine in Asymptomatic Human Immunodeficiency Virus Infection: A Controlled Trial in Persons with Fewer than 500 CD-4 Positive Cells Per Cubic Millimeter," *The New England Journal of Medicine,* **322,** 941–948.

[29]John Moskop, "The Moral Limits to Federal Funding for Kidney Disease," *The Hastings Center Report,* **17,** 11–15.

[30]Enthoven-Kronick, *op. cit.* and Himmelstein-Woolhandler, "A National Health Program for the United States," *op. cit.*

[31]Enthoven-Kronick, *op. cit.,* 29.

[32]*Ibid.,* 31–35.

[33]Himmelstein-Woolhandler, *op. cit.,* 103–106.

[34]Loren Lomasky. (1981) "Medical Progress and National Health Care," *Philosophy and Public Affairs,* **10,** 65–88.

[35]Paul Menzel. (1983) *Medical Costs, Moral Choices: A Philosophy of Health Care Economics in America* (Yale University Press, New Haven, CT); and *Strong Medicine: The Ethical Rationing of Health Care* (Oxford University Press, Oxford, UK, 1990).

[36]Baruch Brody. (1988) "The Macro-Allocation of Health Care Resources," in *Health Care Systems,* Hans-Martin Sass and Robert Massey, eds. (Kluwer Academic Publishers, Dordrecht), 213–236; and "Wholehearted and Halfhearted Care: National Policies vs. Individual Choice," in *Ethical Dimensions of Geriatric Care: Value Conflicts for the 21st Century,* Stuart Spicker, et al., eds. (Kluwer Academic Publishers, Dordrecht), 79–94.

[37]Enthoven-Kronick, *op. cit.,* 32.

[38]*Ibid.*, 35.

[39]Lomasky, *op. cit.*, 69–70.

[40]The most recent advances in genetic engineering and gene therapy are discussed by Ron Kotulak and Peter Gomer in a series of seven articles that appeared in *The Chicago Tribune* from April 8–15, 1990. Each article started on the front page of the paper .

[41]*See* Marcia Angell. (1985) "Cost Containment and the Physician," *Journal of the American Medical Association*, **254**, 1203–1207; The EC/IC Study Group. (1985) "Failure of Extracranial-Intracranial Arterial Bypass to Reduce the Risk of Ischemic Stroke," *The New England Journal of Medicine*, **313**, 1191.

[42]Daniel Callahan. (1990) *What Kind of Life: The Limits of Medical Progress* (Simon and Schuster, New York, NY), 53 .

[43]*Ibid.*, 34.

[44]Menzel offers the most philosophically interesting discussion of what should count as costworthy health care, which he ties in with the "willingness to pay" of individuals. *See* his *Medical Costs, Moral Choices, op. cit.*, chapter one.

[45]Lomasky, *op. cit.*, 75–77.

[46]*Ibid.*, 77.

[47]*Ibid.*, 78.

[48]*Ibid.*, 79.

[49]*Ibid.*, 79–80.

[50]For a good discussion of the issues raised by voluntary health needs, *see* Daniel Wikler. (1987) "Personal Responsibility for Illness," in *Health Care Ethics: An Introduction*, Donald Vandeveer and Tom Regan, eds. (Temple University Press, Philadelphia, PA), 326–358.

[51]Arthur Caplan was one of the first to raise the problems posed by experimental medicine for a national health insurance program. *See* his (1981) "Kidneys, Ethics, and Politics: Policy Lessons of the ESRD Experience," *Journal of Health Politics, Policy, and Law*, **6**, 488–503, at 499–500.

[52]For an excellent and timely discussion of all these issues, *see* Haavi Morreim. (1990) "The New Economics of Medicine: Special Challenges for Psychiatry," *Journal of Medicine and Philosophy*, **15**, 97–119.

[53]John Rawls. (1980) "Kantian Constructivism in Moral Theory," *The Journal of Philosophy*, **72**, 515–572, at 539. *See* also Gerald Winslow. (1986) "Rationing and Publicity," in *The Price of Health*,

George Agich and Charles Begley, eds. (Reidel Publishing, Dordrecht), 199–216.

[54]*See* my paper (1990) "DRGs: Justice and the Invisible Rationing of Health Care Resources," *op. cit.;* and "Justice, HMOs, and the Invisible Rationing of Health Care Resources," *Bioethics,* **4,** 97–120.

[55]For discussion of rationing in Britain, *see* Henry Aaron and William Schwartz. (1984) *The Painful Prescription: Rationing Hospital Care* (The Brookings Institution, Washington, DC); and Thomas Halper. (1989) *The Misfortune of Others: End-Stage Renal Disease in the United Kingdom* (Cambridge University Press, Cambridge, UK). For rationing in America, *see* Victor Fuchs (1984) "The Rationing of Medical Care," *The New England Journal of Medicine,* **311,** 1572–1573; and Lester Thurow. (1984) "Learning to Say 'No'," *The New England Journal of Medicine,* **311,** 1569–1572.

[56]Guido Calabresi and Phillip Bobbitt. (1978), *Tragic Choices* (W.W. Norton, New York, NY).

[57]Baruch Brody, "The Macro-Allocation of Health Care Resources," *op. cit.,* 217.

[58]*Ibid.,* 217–223.

[59]*Ibid.,* 225–226.

[60]*Ibid.,* 227.

[61]*See* Tristram Engelhardt, *The Foundations of Bioethics, op. cit.,* 40–43.

[62]Baruch Brody, *op. cit.,* 228–229.

[63]For a moral justification of the imposition of such controls, *see* Daniel Callahan. (1989) "Rationing Health Care: Will It be Necessary? Can It Be Done Without Age or Disability Discrimination?" *Issues in Law and Medicine,* **5,** 353–366.

[64]This ad appeared in *The Detroit Free Press* (Sept. 14, 1985), their emphasis in the text.

[65]Norman Daniels. (1988) *Am I My Parents' Keeper: An Essay on Justice Between the Young and the Old* (Oxford University Press, Oxford, UK).

[66]Daniel Callahan. (1987), *Setting Limits: Medical Goals in an Aging Society* (Simon and Schuster, New York, NY).

[67]Enthoven-Kronick, *op. cit.,* 94.

[68]Baruch Brody, "The Macro-Allocation of Health Care Resources," *op. cit.,* 213.

Are the NIH Guidelines Adequate for the Care and Protection of Laboratory Animals?

Introduction

In his essay, "Federal Laws and Policies Governing Animal Research—Their History, Nature and Adequacy," Bernard Rollin offers us a brief history of Public Health Service Policy on, and legislation affecting animal rights in experimental contexts. Starting with laws against both blatant cruelty and wanton neglect of animals, the author argues that the practices outlawed by The Animal Welfare Act in 1966 were not motivated by a need to protect animal rights based on the independent moral status of animals in experimental contexts. When first enacted, anticruelty laws were largely irrelevant to the use of animals in scientific research and teaching. In fact, in 1982, the Superior Court of Maryland ruled that anticruelty statutes did not apply in the context of defensible scientific research conducted in private labs. The Animal Welfare Act, as ammended in 1976 was, however, basically an extension of the anticruelty laws to experimental contexts. At any rate, the conservative nature of animal rights legislation before 1985 was clear in that the Congress explicitly disavowed any attempt to regulate research per se conducted in a research facility.

After pointing to the deficiencies of the ammended Animal Welfare Act (which include ignoring animals people do not particularly like, and refusing to encroach on the freedom of scientists in any way), the author points out that the ammended act did not in any way touch on the issues of alleviating experimental suffering or the independent moral worth of animals. Thus, prior to its most recent ammendment in 1985, the Animal Welfare Act cannot be viewed as representing and symbolizing a major advance in social thought regarding the moral status of animals. The act merely reinforced the anticruelty laws on the books and extended them to experimental and educational contexts.

Similarly, Rollin notes that, prior to the mid 1980s, the National Institute of Health's *Guide for Animal Care* represented rules for the care of animals as necessary for the conduct of good science, but even then there was no mechanism for enforcement.

Somewhere in the mid 1980s, however, the animal rights movement began to emerge. Rollin points out that the movement attempted to move the issue of moral concern for, and proper treatment of, animals away from the concepts of kindness and cruelty pivotal to a humane ethic, and to the anticruelty laws. The movement wanted to oblige us, even in scientific contexts, to treat animals properly and to enact legislation in order to codify and protect the independent moral status of animals. The movement implicitly brought with it a higher moral status for animals.

Rollin then seeks to explain how we came to extend basic human rights to animals. This, he claims, is the result of showing that there is no relevant difference between animals and humans that would justify treating animals differently in experimental contexts. Rollin reviews the standard reasons usually given to justify excluding animals from the moral arena as creatures having independent moral worth and whose interests need to be protected even in experimental contexts. For Rollin, all such reasons are defective and, as a result of this position, early legislation focused on eliminating unnecessary pain and suffering. These early laws established positive duties above and beyond avoiding cruelty and maintaining minimum subsistence. Animal-righters thought it weak; researchers thought it too strong.

The two laws establishing such duties were enacted in 1985. The first was the Dole-Brown Bill, which was an amendment to the Animal Welfare Act, and the second was the Health Research Extension Act, which essentially turned NIH policy into law. The remainder of Rollin's essay is a detailed discussion of the relative strengths and weaknesses of both pieces of legislation, and he includes the issue of whether certain research should ever be allowed given the moral worth of the information gleaned by the suffering of animals. For example, should animals be made to suffer in order to certify the safety of various cosmetics? The essay ends with some fertile suggestions as to what sorts of research may not be morally permissible given the independent moral status of the animals involved in the research. He ends with the suggestion that, in science, as elsewhere, animals should be presumed to have a *prima facie* right not to be experimented on, and that we must have good justification in terms of the moral worth of the information to justify taking away that moral right.

In her essay "NIH Guidelines and Animal Welfare," Lilly-Marlene Russow offers an assessment of the current NIH guidelines regarding the care and use of animals in research, education, and testing. After arguing that philosophical discussion on conflicting moral intuitions is not likely to be helpful in resolving fundamentally different perspectives on moral theory as it affects animals, the author describes in detail just what those NIH regulations currently are, and how they affect current resesarch. In the interest of assessing these regulations, one of her basic points is that although the guidelines affirm the independent moral status of animals, the guidelines leave the very concept of the "independent moral status of animals" too nebulous. Much of the essay has to do with the pervasive limitations resulting from this deficiency in the guide. Although the author does not seek explicitly to fully define what is implied by a proper characterization of the "independent moral status of animals," she does point to the kinds of issues that can only be resolved with a proper explication and clarification of that issue.

Along the way, the author discusses issues on the equal moral standing of animals and the degree to which animals deserve equal moral consideration along with humans in research contexts. Although she notes that no philosopher has argued that animals and humans have equal moral standing, it seems clear that the earlier essay by Rollin moves strongly in that direction in its concluding remarks. Also, as was the case in Rollin's essay, Russow discusses at length the question of the moral worth of certain research and whether, given a proper assessment of that worth, one would ever be justified in experimenting on animals. Like Rollin, she stops just short of urging a ban on animal experiments that are likely to cause pain, and that are instigated solely for the purpose of developing products, such as cosmetics, that have little or nothing to do with preventing or curing known diseases.

The author closes the essay with a list of the issues that members of Institutional Animal Care and Use Committees need to consider in the interest of protecting animal rights. They include:

1. a careful assessment of animal needs and welfare along with the ability to predict the impact of various procedures on animals;
2. an understanding of the scientific method as it applies to research involving animals, a committment to the value of such research, and the ability to serve as a respected liaison with the scientific community; and
3. an ability and willingness to think about the appropriate role of scientific research in the larger context of societal values.

Federal Laws and Policies Governing Animal Research

Their History, Nature, and Adequacy

Bernard E. Rollin

Historical Introduction

Though animals have always been viewed as property by the legal system, one can glean some notion of their changing moral status in society by examining the laws that have been, and continue to be, promulgated to assure their welfare. This is true of animals in general as well as, more recently, of animals used in research. Traditionally, beginning in the early nineteenth century, the only protection legally accorded to animals was provided by state laws prohibiting cruelty. The purpose of these laws was to prevent or punish blatant and overt cruelty to animals, thus qualifying to some extent their basic status as property. These laws were aimed at willful, malicious, intentional acts of cruelty, either blatant sadism or wanton negligence, such as not supplying them with food or water, and all standard practices that caused harm to animals in pursuit of human "necessity"—interpreted as economic or other benefit—were excluded from their purview.[1] Thus, common practices in agriculture, rodeo, hunting, trapping,

From: *Biomedical Ethics Reviews • 1990*
Eds.: J. Humber & R. Almeder ©1991 The Humana Press Inc., Clifton, NJ

research, testing, or entertainment were by definition not actionable on grounds of cruelty, however much animal harm and suffering might follow in their wake. As one Colorado judge remarked, these statutes are as much designed to protect the human population from sadists and psychopaths who might begin with animals and escalate to humans as to protect the animals.[2] Since standard practices were, by definition, commonly accepted, and thus, in accord with public morals, they were not covered by these laws. Thus, a New York State judge recently ruled that use of the steel-jawed trap, though in his opinion cruel, was not covered by the anticruelty laws. Any attempt to outlaw the trap, he continued, would require not judicial ruling, but new legislation.[3]

The anticruelty laws were thus largely irrelevant to the use of animals in scientific research and teaching, since these activities were, with a few minor exceptions, either statutorily or judicially exempted from their purview. One notable and well publicized exception is the Taub case, where a researcher who ran a private laboratory, as opposed to operating in a university, was prosecuted successfully in 1982 under the Maryland anticruelty statutes, but was eventually acquitted by Maryland's high court, which affirmed exemption from the law for scientific research.[4]

Such a state of affairs regarding research animals was not offensive to the general public. Animal research was generally perceived to be a necessary part of scientific activity aimed at promoting and augmenting human health, safety, and welfare. Indeed, significant freedom was enjoyed by researchers even in their use of human subjects—so powerful was public confidence in science.[5] The passage of the Laboratory Animal Welfare Act of 1966 by Congress did not mark a revolution in social ethics on animals, nor was it designed to codify any significant change in their moral status. Rather, this law was promulgated in response to sensationalistic press accounts of animals' treatment by those who provided animals to research establishments. The first story concerned the kidnapping of a family dog that was found dead in a research laboratory. The second story was a photographic essay in *Life* magazine that graphically depicted the appalling conditions under which dogs were kept by some animal dealers. The public was both outraged and fearful that their pets might suffer similar fates, and thus, the Congress acted to allay those concerns. It is probably fair to assert that the primary motivation behind the passage of this law was reassuring pet owning humans, rather than protecting animals.

Although the Animal Welfare Act was unquestionably a step forward for laboratory animal welfare, it contained many deficiencies and incoherences. These resulted from two sources: first, the power of the research community, which had no commitment whatever to laws regulating any area of research, and which pushed for as weak a law as possible; and second, the fact that the majority of the public pressure for change was primarily motivated by emotion and sentimental attachment to certain favored animals, and evidenced little understanding of the moral or scientific elements involved in animal research. Even a cursory examination of the 1966 act reveals those difficulties.

The 1966 act concerns itself first and foremost with regulating dealers in dogs and cats to prevent kidnapping of pets and bad treatment of dogs and cats by dealers in response to the publicity mentioned above. Individuals or organizations buying or selling dogs and cats for laboratory use were required to be licensed and were held to certain standards of care and housing for the animals, which standards were to be promulgated by the Secretary of the US Department of Agriculture (USDA), the federal agency charged with enforcing the act. Laboratories using these animals were required to register with the USDA and to identify and keep records on each dog and cat used. In addition, laboratories using dogs and cats were held to USDA standards of care and housing, not only for dogs and cats, but hamsters, guinea pigs, rabbits, and monkeys as well. This law covered only care, treatment, transport, and so on, before and after research, but specifically disavowed any concern with the actual conduct or design of research.

The law was to be enforced by federal inspectors of USDA's APHIS (Animal and Plant Health Inspection Service).

The law was amended in 1970 as a response to continued concern about laboratory animals. These amendments broadened the definition of animal to "any live or dead dog, cat, monkey, guinea pig, hamster, rabbit, or other such warm-blooded animal as the Secretary may determine is being used, or intended for use, for research, testing experimentation, or exhibition purposes, or as a pet." Specifically designated as nonanimals by statute were horses not used for research and any farm animals used in agriculture or agricultural research. Regulations promulgated by the Secretary have thus far continued to exclude rats and mice. Thus, for purposes of the act, a dead dog is an animal, whereas a live mouse is not. The definition of research facility was broadened to include those using animals other than dogs and cats, but the Secretary was permitted to exempt such facilities if they didn't use "substantial numbers" of animals. The definition of dealer was expanded, as was the definition of research facilities. Both civil and criminal penalties were increased for violation of the act (research laboratories are not subject to criminal penalties). The revised law also defined more clearly the sorts of standards to be promulgated by the Secretary—they were to cover housing, food and water, handling, sanitation, ventilation, shelters from extremes of weather and temperature, separation of species, and adequate veterinary care "including the appropriate use of anaesthetic, analgesic or tranquilizing drugs," when such use would be proper in the opinion of the attending veterinarian of research facilities. An annual report must be made to USDA regarding the latter. Detailed regulations were promulgated by the USDA regarding most of the categories specified in the act. The act was again amended in 1976, but these amendments did not affect research, serving primarily to bring common carriers such as airlines under the act, and to cover animal-fighting ventures.

It is clear that the Animal Welfare Act did not represent a serious conceptual augmentation of the social morality about animals underlying the old anticruelty laws. The conservative nature of this legislation prior to 1985 is perhaps best expressed in the strong disavowal of any congressional intention to regulate animal research per se:

Nothing in this Act

(i)...[S]hall be construed as authorizing the Secretary [of Agriculture] to promulgate rules, regulations or orders with regard to the design, outlines, or guidelines of actual research or experimentation by a research facility as determined by such facility.

(ii)...[S]hall be construed as authorizing the Secretary to promulgate rules, regulations, or orders with regard to the performance of actual research or experimentation by a research facility as determined by such research facility.

(iii) [S]hall authorize the Secretary, during inspection, to interrupt the conduct of actual research or experimentation.[6]

The moral limitations of the act were patent. First, the Animal Welfare Act as implemented was quite plain in its concern for favored animals—those deemed desirable by humans—and in its correlative neglect of such nonfavored animals as rodents and farm animals. The Secretary of Agriculture, in drafting the regulations interpreting the act, elected not to include rats, mice, and farm animals used for biomedical research in the scope of enforcing the law, despite the fact that such animals are not statutorily excluded from its purview. Indeed, one may argue that the exclusion of these animals is actually inconsistent with the language of the 1970 amendment; nonetheless, the exclusionary interpretation remains uncorrected.[7]

Second, the act's refusal to encroach on the traditional freedom of scientists in animal research was, as we saw, clear. Third, as an OTA report of 1986 remarked,

Both the legislative and executive commitments of funds and personnel for enforcement have never lived up to the expectations of those who believe the primary mission of the existing law to be the prevention or alleviation of experimental animal suffering.[8]

In sum, then, one cannot regard the Animal Welfare Act, prior to being amended in 1985, as representing and symbolizing a major advance in social thought regarding the moral status of animals. Even the detailed requirements encoded in the regulations drafted by the Secretary may be viewed as minimal standards of food, water, and shelter of the sort that were implicit in the anticruelty laws.

Perhaps surprisingly, the situation was not improved in any meaningful way prior to 1985 by the fact that since 1963, the National Institutes of Health, the major source of funding for biomedical research in the US, had issued guidelines for the care and use of laboratory animals. First published as the *Guide for Laboratory Animal Facilities and Care* and later as the *Guide for the Care and Use of Laboratory Animals,* the document has gone through revisions in 1968, 1972, 1978, and 1985. The provisions of these guides and guidelines did not enjoy statutory authority prior to 1985. Rather, they were allegedly binding on all recipients of NIH funding to do biomedical research in virtue of a contractual agreement entered into with NIH by the researcher and research institution as a condition for the funding. In theory, if the recipient of the funding failed to adhere to these guidelines, all federal research funding to the institution was subject to seizure.

The *Guide* represented principles of husbandry and care of laboratory animals viewed by experts in the field as presuppositional to proper science. Violation of many of these principles would be, in theory, inimical to achieving valid scientific results as well as inimical to the welfare of the animals. Nonetheless, given the historical commitment of NIH to avoid intrusion into the research process, even in the case of research on humans, no mechanism was developed to monitor compliance.

It was widely understood in the research community that NIH did not wish to involve itself in regulatory activities, and thus, its rules and guidelines were often cavalierly ignored. For example, although the 1978 edition of the *Guide* specifically forbade the practice of doing multiple survival surgery on the same animal for teaching surgery, most veterinary schools and many human medical schools openly utilized this practice for economic reasons. Indeed, when I became cognizant of this practice after visiting a number of veterinary schools in the late 1970s and early 1980s, I telephoned NIH and asked why they did not enforce their own rules. "We are not in the enforcement business," a senior official informed me. It was only in the wake of the massive publicity surrounding the major violations of NIH policy exemplified in the Taub case (1982), that NIH exercised for the first time its power to seize funding. Following the Taub case and the release shortly thereafter of the infamous University of Pennsylvania head-injury laboratory videotapes, which again documented flagrant violations of NIH policy, NIH at last moved to ensure institutional compliance with its policies. Funding was made available to NIH's Office of Protection from Research Risks to make both unannounced and announced site visits to ensure compliance. Subsequent to the implementation of this policy, a number of institutions were found deficient and funding was indeed seized.

The Emergence
of a New Social Ethic for Animals

One can see from the above discussion that moral concern for research animals was not a major legal or social priority for most of this century. Gradually, however, this began to change with the advent of a new social ethic articulating human obligations to animals. Though difficult to date with any precision, this ethic had been transmuted into a significant social and political force by the mid 1980s. For want of a better term, and for the sake of brevity, one may characterize this new movement in thought and action as the "animal rights movement," though, as in any social movement, one rubric does not do justice to the diversity of opinions it attempts to subsume. For example, many of the philosophers articulating the key ideas of the movement, most notably Peter Singer, have strong ultimate philosophical reservations about the cogency of the concept of rights.[9] Such differences not withstanding, however, one can usefully sketch some major features of this new movement:

In the first place, it seems clear that the intellectual roots of the movement lie in the 1960s. The characteristic 1960s concern with extending moral and legal protection to the disenfranchised—blacks, women, minorities, children—was adopted quite directly by the animal movement, and these issues were consciously tied to animal issues.

Second, the new movement, in its thought, consciously attempted to move the issue of moral concern for and proper treatment of animals away from the concepts of kindness and cruelty pivotal to the traditional humane ethic, and to the anticruelty laws. In this new thinking, humaneness is replaced by justice; we are obliged to treat animals properly—doing so is not merely an overflowing of benevolence.

Third, the animal movement benefitted from the rise of public disenchantment with science and technology that began in the 1960s, as well as from the rise of public concern with the nonhuman world through the environmental movement. Though there are deep and ultimate differences between the two movements, superficially, they both reinforce one another by putting ethical emphasis on the extra-human features of the world.

Fourth, and most relevant to our discussion, the new movement pushed for new legislation to codify and protect the moral status of animals. We earlier alluded to the New York State judge who suggested that eliminating the steel-jawed trap could only be effected by new legislation. And this strategy is precisely what the animal movement worldwide has attempted to achieve, primarily in the area of animal research but also in the area of animal agriculture.

One of the great strengths of the animal movement has been its emphasis on a sound philosophical base. Numerous philosophers, though differing as we said in many details, have nonetheless helped construct a powerful argument that undergirds the movement's political activities. The argument does not so much recommend a new ethic for the treatment of animals as suggest that a higher moral status for animals is logically implicit in the consensus ethic for humans we already share in democratic societies, and just needs to be brought forth through dialectically examining that ethic.

Let us, therefore, briefly summarize the socially accepted ethical ideal for humans that pervades our thinking and practice before we explain the attempt to apply it to animals.[10] In democratic societies, we accept the notion that individual humans are the basic objects of moral concern, not the state, the Reich, the Volk, the Church, or some other abstract entity. We attempt to cash out this insight in part by generally making our social decisions in terms of what would benefit the majority, the preponderance of individuals, i.e., in utilitarian terms of greatest benefit to the greatest number. In such calculations, each individual is counted as one, and thus, no one's interests are ignored. But such decision-making presents the risk of riding roughshod over the minority in any given case. So democratic societies have developed the notion of individual rights, protective fences built around the individual that guard him or her in certain ways from encroachment by the interests of the majority.

These rights are based on plausible hypotheses about human nature, i.e., about the interests or needs of human beings that are central to people, and whose infringement or thwarting matters most to people (or, we feel, ought to matter). So, for example, we protect freedom of speech, even when virtually no one wishes to hear the speaker's ideas, say in the case of a Nazi, regardless of cost. Similarly, we protect the right of assembly, choosing one's own companions, one's own beliefs, and also the individual's right not to be tortured, even if it is in the general interest to torture, as in the case of a criminal who has stolen and hidden vast amounts of public money. And all of these rights are not simply abstract moral notions, but are built into the legal system. Thus, the notion of human nature is pivotal to our ethic—we feel obliged to protect the set of needs and desires that we hypothesize as being at the core of what it means to be human.

What we have so far outlined is not difficult to extract from most people in our society. And, in fact, calling attention to the moral principles people unconsciously accept has traditionally been a major way of effecting social change. Arguably, something of this sort occurred when (thinking) segregationists accepted integration, or when occupations such as veterinary medicine, which had traditionally barred women, began to admit them. In both of these cases, presumably no change in moral principles is required. What is demanded is a realization that a moral commitment to equality of opportunity, justice, fairness, and so on, that the segregationists or persons who barred women from veterinary school themselves had as a fundamental commitment, entails a change in practice. In other words, such people had readily accepted democratic moral principles as applying to all persons. What they had ignored was the fact that the class of persons was far greater than what they acknowledged, and included blacks and women. And it is this same sharpening of the applicability of accepted moral principles, rather than the adoption of new ones, that led to the steady augmentation of the class of full rights bearers in the US Constitution beyond the original limited group of white, adult, native-born, male, property owners.

As we all know, the history of western civilization in general, and western democracies in particular, has been one of ever-increasing extension of moral concern to previously disenfranchised individuals. Whereas at various points in history individuals were excluded from moral and legal protection, or from having their interests compete in the moral arena on the basis of characteristics like ancestry, gender, color, religion, age, race, citizenry, or nationality, such traits have gradually come to be seen as lacking moral relevance. In other words, although there are, of course, differences between whites and blacks, men and women, citizens and foreigners, the key question has been: Do these differences justify a difference in how they ought to be treated morally. The answer has been negative.

The next crucial step is to show how the above ethic can inexorably be applied to animals as well. If one can show that there are no rationally defensible grounds for differentiating animals from humans as candidates for moral concern, we must logically bring to bear on questions of animal treatment the entire moral machinery we use to deal with human questions. And it turns out that none of the standard reasons offered up in the history of thought to exclude animals from the moral arena will stand up to rational scrutiny and meet the test of moral relevance.

For example, it has often been suggested that animals are not fully worthy of moral concern because they don't have immortal souls (incidentally, one is surprised at how often scientists say something like this; witness the letters in US veterinary journals). Aside from our obvious inability to state with any certainty who does or doesn't have a soul or even what it is, this view is open to a much more striking response, first enunciated by Cardinal Bellarmine. True, said Bellarmine, as a Catholic, I must accept Church doctrine that animals have no souls, but this has nothing to do with their being excluded from moral concern. In fact, he argued, they ought to be treated better than people, since this is their only chance at existence, whereas wrongs to humans will be redressed in the afterlife! So the really interesting point in this example is not theological, but rather the dramatic way it enjoins us to be clear about the moral relevance of the differences we point to.

Other alleged differences between people and animals fare no better. Some have said we can do as we wish to animals because we are superior; but what does "superior" mean? It is sometimes claimed that it means that we are "at the top of the evolutionary ladder" but, as we all know, there is no evolutionary ladder, only a branching tree. And if it does make sense to talk about species superiority, it is only in terms of differential reproduction, species longevity, and adaptability, in which case we share top billing with many other species, like the rat, and both lose hands down to the cockroach. If "superior" means that we are more powerful than other creatures and can in fact do as we wish with them, this is surely true, but has no moral relevance. To say that it does is to affirm that might makes right and to destroy morality altogether, to confuse *de facto* authority with *de jure* authority. If one accepts this position, one is forced to say that the government has the right to kill people as it sees fit, since it is, after all, more powerful than any of us, or that the mugger or rapist is perfectly morally justified in exploiting his victim.

Still other alleged differences turn out to be equally irrelevant from a moral point of view. "Man is rational, animals are not," is a favorite justification for the exclusion of animals from moral concern. But what is the moral relevance of rationality? Doubtless one needs to be a rational being to be a moral agent or actor, to be held morally responsible for what one does. But one surely doesn't need to be rational to be an object of moral attention and concern—consider children, infants, the insane, the senile, the comatose, the retarded, and so on. And furthermore, if rationality is the key feature of what makes something worthy of moral attention, one may ask why so much of our moral concern is devoted to aspects of human life that have nothing to do with rationality? Suppose I discovered that I could make my students more rational by wiring their seats and shocking them when their attention wandered. If rationality were the key feature relevant to moral concern, such behavior would not only be permissible, but obligatory. Yet we would rightly condemn such behavior as monstrous, showing that rationality is not the only thing involved in being an object of moral concern. There are many other features involved, as indeed this hypothetical case shows. It is wrong to shock the students because it causes pain and infringes on their freedom. Various other differences, such as the claim that ethics is based in social contracts and animals can't enter into contracts, turn out to be either false or to lack the requisite degree of moral relevance that would justify not considering them morally.

Equally important, a moment's reflection makes it patent that not only are there no morally relevant differences for excluding animals from moral concern as we in society define it, but there are, in fact, significant morally relevant similarities that animals share with humans. The same sorts of features that we find in people, which give rise to our talking about right and wrong actions with regard to people, are also to be found in animals. The features I am talking about that are common to people and to at least "higher" animals (and possibly many "lower" ones as well) are interests—needs, desires, predilections, the fulfillment and thwarting of which matter to the person or animal in question. Cars have needs—for gas, oil, and so on—but they do not have interests, since we have absolutely no reason to believe that it matters to the car itself whether or not it gets its oil. That is why it is impossible to behave immorally towards cars in themselves—they are merely tools for human benefit. But animals with interests cannot be looked at as mere tools, for they have lives that matter to them.

There are, of course, categories of interests and interests that are common to all animals (including humans)—food, reproduction, avoidance of pain. But even more significant are the unique variation on these general interests, and the particular interests, that arise in different species. Even as we talk of human nature, as defined by the particular set of interests constitutive of and fundamental to the human animal, we can also talk of animal natures as well—the "pigness" of the pig, the "dogness" of the dog. Following Aristotle, I like to talk of the *telos* of different species of animals as being the distinctive set of needs and interests, physical and behavioral, genetically determined and environmentally expressed, that determine the sort of life it is suited to live. This is not a mystical notion—it follows directly from modern biology and genetics, and is certainly obvious to anyone who is around animals and indeed to common sense; hence, the song that tells us that "fish gotta swim and birds gotta fly."

Recall that we have argued that our consensus ethic for humans protects certain aspects of human nature deemed to be essential to the human *telos* and shields them from infringement by the majority and by the general welfare. If it is the case that one can find no morally relevant grounds for excluding animals from the application of that ethic and if animals too have a *telos,* it follows inexorably that animals too should have their fundamental interests encoded in and protected by rights that enjoy both a legal and moral status. In this way, we indeed illustrate that the notion of animal rights is implicit (albeit unrecognized) in our consensus social ethics.

It is unquestionable that something like this ethical ideal has developed internationally—it is, for example, the backbone of new Swedish legislation, which guts confinement agriculture and mandates that animals be raised in such a way as to accommodate their natures.[11] Patently, with the emergence of this ideal, traditional scientific *laissez-faire* regarding the use of animals was no longer acceptable. The stage was set for new legislative constraints on animal research.

The Emergence of New Legislation

The attempt to translate this sort of ideal into meaningful, politically viable legislation aimed at improving the lot of research animals began in the mid 1970s. At that time, a Colorado-based group consisting of researchers, attorneys, and myself (a philosopher) spent a number of years drafting legislation that would move society a step closer to the ideal.[12] Paramount to our concerns was the notion that it was at least incumbent on the research community to eliminate pain and suffering in research that was not essential to what was being studied, and to maximize the interests of animals used in research consonant with that research. Such a general notion, we felt, was best monitored by local committees of scientists and lay persons and animal advocates. Though quite a weak notion, such a concept nonetheless represented a major step beyond the traditional emphasis on cruelty and beyond the established Animal Welfare Act, in that it entailed that those using animals had positive duties to those animals beyond providing minimum subsistence and avoiding cruelty.

Predictably, much of the animal rights community perceived the concept as far too weak; at the same time, the research community perceived it as unacceptably restrictive, given the history of total freedom in animal use. But by 1982, when I was called to testify on behalf of the so-called Walgren version of the bill, I carried the active support of most of the mainstream animal organizations as well as that of some major portions of the research community, including the American Physiological Society, the traditional opponent of any intrusion into the research process. By 1985, our concepts had been encoded in two pieces of legislation, one an amendment to the Animal Welfare Act known as the Dole-Brown Bill, the other, a bill essentially turning NIH Policy into law, known as the Health Research Extension Act. Before discussing the strengths and limitations of each of these pieces of legislation, it is essential that we summarize them. The major statutory provisions of the Amendment to the Animal Welfare Act (PL 99–198) are as follows:

1. Establishment of an institutional animal care committee to monitor animal care and inspect facilities. Members must include a veterinarian and a person not affiliated with the research facility.
2. Standards for exercise of dogs are to be promulgated by the Secretary of Agriculture.
3. Standards for a physical environment that promotes "the psychological well-being of primates" are to be promulgated.

4. Standards for adequate veterinary care, including use of anesthetics, analgesics, and tranquilizers, are to be promulgated.
5. No paralytics are to be used without anesthetics.
6. Alternatives to painful procedures must be considered by the investigator.
7. Multiple surgery is prohibited except for "scientific necessity."
8. The Animal Care Committee must inspect all facilities semiannually, review practices involving pain, review the conditions of animals, and file an inspection report detailing violations and deficiencies. Minority reports must also be filed.
9. The Secretary is directed to establish an information service at the National Agricultural Library that provides

laissez-faire in research and now did propose to regulate certain aspects of the design and conduct of research, most notably that relevant to the control of pain and suffering, the drafting of these regulations involved considerable controversy. Although the law passed in 1985, parts 1 and 2 of the regulations were not finalized until late 1989, and the USDA received over 7,000 comments concerning the proposed regulations. The controversy over Part 3, that defining the requirement of exercise for dogs and environments for primates that enhance their psychological well-being, continues to rage (as of late 1989), with major portions of the research community arguing that psychological well-being is scientifically meaningless. As a result, the regulations interpreting these provisions are not expected to be released until late in 1990.

One major focus of the regulations is an attempt to harmonize the requirements of the Animal Welfare Act with that of the NIH policy, which, as we mentioned, was made into law in 1985 by the Health Research Extension Act (PL 99–158). This act essentially puts the power of law behind the NIH *Guide* and applies to any institution receiving NIH money as well as to NIH's own intramural laboratories. Each grantee must submit an assurance statement to NIH that they will comply with NIH policy. Violation of such policy entails seizure of funding. In addition, grantees must have an Institutional Animal Care and Use Committee (IACUC) that must meet regularly and review research and teaching protocols involving animals before they are funded; inspects facilities at least semiannually; assures that all researchers, technicians, and other personnel involved with animals are properly trained; and that may suspend any activity involving animals not in compliance with policy. The committee must consist of at least five members: one, a veterinarian with laboratory animal medicine background; one a scientist experienced in animal research; one member whose primary expertise is in a nonscientific area; and one public member not affiliated with the institution. In September of 1986, the Office of Protection from

Research Risks of NIH issued a detailed account of its regulations interpreting the new law, entitled the *Public Health Service Policy on Humane Care and Use of Laboratory Animals*. Among other provisions, the new rules require that institutions designate clear lines of authority and responsibility in animal care and use. Record keeping requirements have also been strengthened. In addition, the new *Guide* has strengthened the requirements for adequate veterinary care, makes reference to the social environment for laboratory animals, requires aseptic surgery for rodents, and strengthens euthanasia requirements. Animals that experience pain that cannot be alleviated must be euthanized at the end of or during an experimental procedure, in accordance with the recommendations for the American Veterinary Medical Association Panel on euthanasia.

The two new laws, taken in tandem, assure that the majority of animals used in research in the US fall under some mandated protection. Unlike the Animal Welfare Act, the NIH policy does not exclude rats, mice, or farm animals used in biomedical research. Committees must review facilities and care for all animals used in biomedical research and review protocols using all animals. As in the 1985 amendments to the Animal Welfare Act, major emphasis is placed by committees on control of pain and suffering. Most committees have interpreted the NIH law, which statutorily specifies that it applies to "behavioral and biomedical research," as applying to field research in wildlife and zoological research as well as biomedical, thus protecting an entire range of animals not traditionally thought of as research animals.

A final strengthening of federal animal welfare policy is embodied in a 1983 memorandum of understanding between APHIS, NIH, and FDA, agreeing to share information on institutions visited by each agency. In effect, this memorandum empowers the USDA inspectors, who regularly inspect institutions, to report violations of NIH policy to NIH, even if such violation concerns areas or animals not covered by the Animal Welfare Act.

Positive Features of the New Laws

To adequately assess the positive features of the new laws requires some grounding in what I have elsewhere called the common sense or ideology of twentieth century science.[13]

Although twentieth century science has tended, quite intentionally, to separate itself from philosophical concerns, it is patent that no area of human activity can avoid making philosophical commitments, for all disciplines must rest on concepts and assumptions taken for granted by practitioners of the discipline. Twentieth century science, too, has its philosophy, though that philosophy is typically invisible to its practitioners, who tend to see science's assumptions not as debatable philosophical precepts, but as self-evident truths. Thus, the philosophical assumptions made by science include an aversion to philosophical examination of these assumptions, and, in part for that reason, they have tended to harden into an ideology virtually universally pervasive among scientists.

One major component of scientific common sense, directly relevant to the issue of animal use in biomedicine, is the belief that science is value-free, ought to make no valuational commitments, and thus, *a fortiori*, has no truck with ethics. This notion, like many other components of scientific common sense, is rooted in the logical positivism of the early twentieth century, which stressed the need for objectivity, empiricism, and verification in science. Since value claims in general, and ethical claims in particular, are not subject to empirical test and verification, they have no place in science. They are at best, to scientific ideology, emotional predilections and cannot be dealt with objectively. It is for this reason that otherwise cool and rational scientists are often every bit as emotional on such ethical issues as animal use as their opponents are—their training and ideology has led them to the view that ethical issues are, in fact, nothing but emotional issues, where rational thought has no place, and they thus believe that battles are won by manipulating emotions and tugging at heartstrings. The possibility of a rational ethic on anything is instinctively seen as an oxymoron or solecism.

It is not difficult to determine that science has indeed viewed itself as value-free. Introductory textbooks, such as Keeton's or Mader's recent college biology texts, stress in their preliminary discussions of science and scientific methodology that science is value-free and unable to make ethical pronouncements. Such a position is reflected in the teaching of science, where science educators typically make such a doctrine explicit, or implicitly communicate it by their failure to discuss ethical issues occasioned by the material they are teaching. Leading scientists, in public pronouncements, promulgate the value-free view of scientific inquiry. Thus, in 1989, James Wyngaarden, former Director of NIH, affirmed that research into genetic engineering should not be hindered by ethical concerns.[14] Another excellent example of this may be found in a recent PBS television documentary dealing with the Manhattan Project, the development of the atomic bomb during World War II. When queried as to their ethical stance on the development of the bomb, most of the scientists replied that they left such questions to the politicians since ethics is not in the purview of scientists. Scientific journals rarely articulate ethical issues occasioned by their subject matter, and scientific conferences consider them only when the issues are galvanizing significant concern among society at large.

It is not surprising, therefore, that it has been widespread among scientists to assume that the use of animals in research is simply a scientific question to which ethics is irrelevant. An excellent paradigmatic example of this view may be found in a recent textbook of abnormal psychology authored by two distinguished researchers. In the book, there is a photograph of a laboratory rat, accompanied by the following caption: "For moral reasons, animals are used in psychological research."[15] This statement typifies the idea that, in working with animals, one has somehow circumvented the need for dealing with moral issues.

A moment's reflection reveals that the ideological view of science as value-free in general, or ethics-free in particular, must be wrong.

Indeed, one need look no further than the issue at hand, the use of animals in research, to find a paradigmatic example of a moral judgment built into the very foundation of scientific activity. Insofar as we forebear from doing biomedical research on unwanted children or political prisoners, even though they are scientifically a far higher fidelity model for the rest of us than rats or other animals are, we have a moral judgment standing at the very basis of biomedical science.

For that matter, let us recall that many scientists argue that modern biomedicine is essentially connected to and dependent on the invasive use of animals in order to function and progress. Connected with this claim, then, is a series of assumptions made by researchers, namely that the advance of applied knowledge that benefits humans, or for that matter, of pure knowledge with no obvious use, licenses the invasive use of animals or that the knowledge or control gained through research is worth more than the pain or death of the animals used in that research process. Such judgments are widely shared, perhaps universally shared among researchers, but they are unquestionably moral judgments.

One can see immediately that the new federal laws entail a significant and frontal attack on this "value-free" aspect of scientific ideology. For both laws make quite clear that animals enjoy a moral status beyond that of mere tools for research and one that ensures greater protection than that encoded in traditional prohibitions against cruelty. One can indeed argue that the new laws encode some limited rights for animals in the sense discussed above—research animals are now entitled to the alleviation of pain and suffering not essential to the research in question, even if such alleviation is expensive or burdensome to the researcher. In other words, in a limited way, the moral status of animals trumps human utility. And insofar as the Amendment to the Animal Welfare Act requires exercise for dogs and enriched environments for primates, it again codifies some rights for a limited set of animals.

A second way in which the new laws erode science's distancing of ethics is the requirement for review of protocols. One can hardly discuss and evaluate protocols for animal use without coming up against a whole host of moral and morally tinged questions—can and ought the same ends be achieved without inflicting pain and suffering; has everything been done to assure the animals' comfort; does the knowledge gained always (or ever) justify the animals' suffering; should painless euthanasia be construed as harming an animal; does the fact that animals suffer in the wild have any moral relevance to the fact that we inflict suffering on them for research; do we have a right to hurt animals for our benefit? For their benefit? Even though some of these questions are legally mandated and others are not, all of the above and more inevitably arise in open discussions.[16] And even though, theoretically, animal care committees are not mandated to look at and weigh cost-benefit (i.e., suffering of the animals versus benefit to humans, science, or other animals) the way human subjects committees do, or at scientific merit of proposals, but only at alleviation of suffering, in practice the distinction is impossible to maintain. Many times, members of committees will feel that a procedure, though state-of-the-art scientifically, should give way to something less invasive. Other times, especially when looking at projects that are not funded by NIH or other agencies and thus are not peer-reviewed, committees may indeed use a cost-benefit standard. This is, in fact, almost impossible to avoid when looking at teaching uses of animals. Furthermore, insofar as the laws expect committees to look at whether the researcher is using the proper species and number of animals, the distinction between scientific and welfare considerations is further eroded.

Thus, a major benefit of the new laws is to generate among scientists dialogue on morally problematic aspects of animal research, thereby eroding the indefensible ideology that, in the past, served to allow animal research to be seen as a value-free activity.

Yet another pernicious aspect of scientific ideology is eroded by the new laws: As I have discussed at length elsewhere,[17] one component of scientific ideology has been the claim that one cannot assert, legitimately, that animals are conscious in the sense of enjoying subjective experiences, feeling pain, fear, anxiety, loneliness, boredom, joy, happiness, pleasure, and the other noxious and positive mental states that figure so significantly in our moral concern for humans. This skepticism about attributing thought to animals enjoys a long history, and was most famously promulgated by Descartes, who declared that animals were simply machines, driven by clockwork. Such a position, of course, provided justification for experiments in the burgeoning science of physiology in Descartes' time that required dissection of living animals without anesthesia. While Descartes' position was hotly contested by many philosophers and scientists, most notably by Darwin, who argued that if physiological and morphological traits were phylogenetically continuous, so too were mental and psychological ones, agnosticism about animal minds resurfaced in the early 20th Century, receiving succor from non-Cartesian sources. We mentioned earlier that the positivism that shaped scientific ideology denied the validity of talking about ethics, since moral claims were not verifiable. The same positivistic tendency nurtured the development of psychological behaviorism, which denied the studiability of mind and consciousness, and affirmed that only overt behavior was open to scientific inquiry. This methodological aversion to treating mental states as real was enormously influential, shaping the thinking of psychologists, zoologists, biomedical scientists, and even the European ethicists who otherwise rejected behaviorism.

It is clear that the denial of mentation to animals did have untoward moral consequences in science. Scientific books and papers routinely stopped short of attributing felt pain, fear, and so on to animals, and any such extrapolations beyond overt behavior were seen as pernicious "anthropomorphism," this despite the fact that much animal research, for example pain research, presupposed that animals could feel pain. Though all analgesics in the US were routinely tested on laboratory animals, these animals virtually never received analgesics in the course of research, and one searched in vain for a literature on laboratory animal analgesia. Incredibly, the first conference on animal pain ever held in the US was only convened in 1983, and, even then, dealt almost exclusively with the machinery or "plumbing" of pain, ignoring the subjective and morally relevant aspects.[18] The scientific literature never discussed suffering in animals, and in its zeal to avoid "unverifiable" talk about mental states like fear, anxiety, loneliness, boredom, the research community talked blanketly in terms of mechanical, physiological "stress responses," that tended to be simplistically defined in terms of Cannon's alarm reaction for short-term stress, or Selye's activation of the pituitary adrenal axis for long-term stress.

One of the major salubrious effects of the new laws has been their emphasis on animal pain, suffering, and distress, and how these are to be controlled. In the face of public law that affirms that animals feel pain, fear, distress, boredom, loneliness, and so on, the research community cannot maintain its unjustified agnosticism about these states. I say unjustified because it is my contention that science cannot avoid talking about mental states in animals, that failure to do so makes for inadequate explanation, and that such talk is no more unscientific than a whole host of other accepted scientific notions. We cannot verify any of the following directly experientially: the existence of an external world independent of our perceptions; the reality of the past; or the existence of quanta; yet all of these are presupposed in science because they provide us with vehicles of explanation that generate testable consequences. The same point held of mental states in animals—they help us to explain how the animals behave. Recent research has shown, for example, that talking about stress in animals cannot be done in purely mechanical terms and cannot circumvent reference to their states of awareness, for the same "stressor" can have very different physiological effects in an animal depending on the animal's emotional state, or on how it has been treated prior to the noxious stimulus, or on whether it can anticipate or control the stimulus, and so on.[19] Furthermore, purely psychological stressors, like putting an animal into an unfamiliar environment, can have greater physiological effects than such "physical" stressors as heat.[20] By the same token, pain as an experienced state in animals must be postulated for a variety of compelling reasons—for one thing, to make pain research on animals a coherent project! The mechanisms of pain activation and pain mitigation are virtually the same in all vertebrates (e.g., endorphins and enkephalins, serotonin, substance P, stimulation produced analgesia),[21] and if pain were purely mechanical and unfelt in animals, why would the experiential dimension suddenly emerge evolutionarily in humans (we know, in fact, that humans who have congenital or acquired inability to feel pain do not do well at all)?[22]

As soon as the laws passed, scientists were forced to shed their ideological skepticism about animal consciousness and "reappropriate common sense." Not surprisingly, more papers on the recognition, control, and alleviation of animal pain have been published since the passage of the new laws than were published during the previous hundred years! Whereas, prior to the laws, one could distance oneself from animal pain and suffering behind ideological barriers, one must now meet it—and deal with it—head-on.

In sum then, the new laws have begun to erode the entrenched scientific ideology that asserts that science has no truck with ethics (and with the ethical issues raised by animal research), and that one cannot (and, therefore, need not) deal with or even acknowledge the existence of animal pain, suffering and other modalities of consciousness (such as "psychological well being") in science. Thus, genuine and major efforts to control pain and suffering and develop less invasive alternatives to traditional practices have been directly occasioned by the new laws, as has greater attention to animal mentation in general.[23]

Another positive feature of the new laws is the emphasis placed on training of researchers and other personnel in proper care and use of animals and in such invasive techniques as surgery. In the past, one could get an MD-PhD degree from a major university in an animal-using area of biomedicine and never learn anything about the animals one uses other than that they model some particular disease and syndrome. Such ignorance led not only to bad science resulting from failure to control for relevant metabolic and physiological variables, but very obviously to animal pain and suffering growing not out of malice, but ignorance.

Another area in which the laws have had a very positive effect is their insistence on clear systems of authority, responsibility, and accountability in animal care. All too often in the past, care was left in the hands of investigators and graduate students, and assumed a very low priority in the face of weekends, holidays, deadlines, and money shortages. The centralization of animal care, or at least of responsibility for animal care, goes a long way towards eliminating suffering arising from neglect.

Finally, built into the Animal Welfare Act amendments is the seed of the idea that animals need more than just food, water, bedding and painkillers to live morally acceptable lives. Other aspects of their *telos* must also be respected. Indeed, one NIH official confided to me the opinion that more animal suffering results from our housing and husbandry of animals under conditions convenient to us and not congenial to their natures than arises out of invasive manipulations. Whether this is the case or not, it is plain that research animals would be better off if we acknowledged their *telos* in our husbandry systems. The new law, in its requirement of exercise for dogs and psychological well-being of primates, blazes a small but significant trail into this hitherto uncharted area.

There is, in sum, no question that the new laws have given rise to greater researcher sensitivity to animal pain, suffering and needs as well as flagged the need for implementation of the classical 3 Rs of alternatives to animal use—reduction in the numbers of animals used, refinement in manipulative techniques, and replacement of animals in research by other techniques.[24]

They have also focused emphasis on much needed training, and have spurred conferences and papers on animal analgesia, stress, pain, and so on. They have also forced scientists to start thinking about ethical issues associated with animal use. Hopefully, they will breed a generation of scientists to whom these concerns are second nature, and who have transcended the ideology we discussed earlier.

Limitations and Inadequacies in the New Laws

Despite these salubrious developments, the new laws are by no means ideal or even totally adequate. In the first place, not all animals used in research are covered. Neither of the new laws applies to rats and mice, farm animals, or birds used in industry, since the Animal Welfare Act still doesn't consider these creatures to be animals, and the NIH law only applies to federal grant recipients or NIH's own labs. Clearly, the Animal Welfare Act must be extended to cover all animals; as we saw earlier, the exclusion of rats, mice, birds and farm animals used in biomedicine by the Secretary actually seems to go against the intent of the law.

In addition, these laws are currently restricted in their application either to warm-blooded animals (Animal Welfare Act) or to vertebrates (NIH law). These cut-off points are clearly arbitrary, and many committees have, to their credit, extended their application to such higher invertebrates as squids, where there are good scientific reasons to suspect the presence of thought and feeling. The scope of these laws should be statutorily expanded to include all animals where there are good reasons to infer the presence of pain and/or consciousness.

Another marked inadequacy in these laws pertains to animals used in agricultural research. The Animal Welfare Act spe-

cifically excludes from its purview farm animals used in agricultural research; NIH policy too, does not apply to farm animals
used in agricultural research. Yet millions of farm animals are
used in such research in ways that may be as invasive and occasion
as much pain and suffering as biomedical research. Such agricultural projects may include surgery, deficient diets, food and water
deprivation, total confinement, and induced disease, yet these
animals enjoy no legal protection. Thus, suppose one has twin male
lambs, one of which goes to a NIH-funded biomedical research
project, the second to an agricultural research project. Both are to
be castrated. The NIH lamb will get anesthesia, post-surgical
analgesia, and will be castrated under aseptic conditions. The agricultural lamb may have the testicles removed under field conditions
in standard ways—which include having them bitten off!

To their credit, many committees now apply NIH standards
to all surgical procedures done on their campuses—even for agricultural research. Nonetheless, agricultural animals clearly need
to be included under standards as rigorous as those governing the
treatment of biomedical animals. The agricultural research
community has recently adopted voluntary guidelines for their
research animals; not surprisingly, these are both far too weak and
have no enforcement structure to back them.

The major criticisms of these new laws, however, stem from
the fact that they don't go far enough. Some philosophers have
made the case that ultimately there is no moral justification for
invasively using animals in research at all.[25] When researchers
attempt to answer this sort of argument, they respond in cost-
benefit terms—that the good to humans and animals coming out
of research outweighs the cost in pain and suffering to the research animals. Leaving aside the cogency of this response, one
can indeed acknowledge that this statement seems to capture the
current state of social moral thought on this question. But if this
is indeed the case, then it naturally follows that the only invasive
research that ought to be pursued is research where the benefit to
humans and/or animals likely to emerge from the research out-

weighs the cost in suffering to the animals. I have elsewhere called this the Utilitarian Principle.[26]

This maxim suggests that much invasive research, which is aimed at "pure knowledge," should not be allowed. One standard researcher response to this principle is to invoke the serendipity argument. It is argued that though it may not appear that a particular piece of research will produce foreseeable benefit, one never knows what will arise adventitiously. The response to that is simple: By definition, one cannot plan for serendipity. Society does not fund a great deal of research for a wide variety of reasons. Much research is turned down by the granting agencies because it is perceived as poorly designed, less important than other things, and so on. If the serendipity argument were valid, one could not make such discriminations, and one would be logically compelled to fund everything.

Admittedly such cost-benefit calculation as we suggest is fraught with difficulties—how does one weigh one parameter against a disparate one? But the crucial point to remember is that we do currently make such cost-benefit decisions in a variety of areas, including research on humans. All that needs be done is that such calculations be exported to the area of animal use. Certainly, there will be hard cases, but at least extreme cases will be clear. Invasive research aimed at developing a new weapon, a new nail-polish, or at discovering knowledge of no clear benefit to humans and/or animals, for example territorial aggression studies, would clearly not be permitted.

Obviously then, some mechanism needs to be developed that will exclude invasive research that produces no benefit, but simply advances knowledge or careers. Some types of psychological research, for example, are very vulnerable to this criticism. The current mechanism of peer review, whereby experts in the field judge the value and fundability of research plainly does not address these concerns. Researchers who throughout their whole careers have taken a particular sort of invasive animal use for granted in their field are not the best source for eliminating such a use from the field. A better alternative, perhaps, would be to allow local committees with greater representation from the citizenry at large to pass on the value of a piece of animal research. Society pays for animal research; researchers ought to be able to successfully defend their need to spend public money to hurt animals to a set of citizens. Such an approach works for our justice system; perhaps researchers need to convince something comparable to a jury of their need to hurt animals for the sake of research.

Thus, I would argue that local committees should also be charged with deciding whether a piece of research ought to be done, and that such committees be made up of a majority of non-scientists representing the public in general.

The final major area of deficiency has already been alluded to earlier in our discussion. When we discussed the emerging social ethic for animals, we argued that pivotal to that ethic is the need to protect certain fundamental aspects of the animals natures. The new laws focus on pain and suffering growing out of research use, and only begin to touch on deprivation, growing out of husbandry and housing, in the case of dogs and primates. I would argue that the ethic when applied to research animals demands another principle, which I have elsewhere called the Rights Principle,[27] which asserts that in the context of research, all research should be conducted in such a way as to maximize the animal's potential for living its life according to its nature or *telos*, and certain fundamental rights should be preserved regardless of considerations of cost. In other words, if we are embarking on a piece of research that meets the Utilitarian Principle, we by no means have *carte blanche,* we must attend to the animal's rights following from its nature—the right to be free from pain, to be housed and fed in accordance with its nature, to exercise, to companionship it is a social animal, and so on. The animal used in research should thus be treated, in Kant's terminology, as an end in itself, not merely as a means or tool.

I would, therefore, argue that the laws mandate the creation of husbandry and housing systems that allow the animals to live lives approximating that dictated by their *telos,* so as to assure as much as possible their happiness, as well as the mitigation of their pain and suffering. Precedent for this already exists in the work done on enriched environment for zoo animals.

In sum, the new laws have taken significant steps to augment and protect the moral status of animals, and do reflect the growing ethical ideal for animals emerging in society that we summarized earlier. The fact that they have drawn opprobrium from both extremes bespeaks their adherence to mainstream social thought. All social revolutions in the US have proceeded incrementally, and probably none have been as radically new and unprecedented as according rights to animals.

At the same time as we have praised the laws for elevating the status of animals and helping to erode pernicious scientific ideology, we must also realize that they are just a beginning. Happily, they contain room for growth and development, largely through their emphasis on enforced, local self-regulation. Despite the variegated natures of different committees at different institutions, national consensus standards do seem to emerge, as do solutions to moral problems that arise at one institution and become appropriated elsewhere. Thus, there is room for the committee structure to both shape and reflect changing social thought on animals, for example, by being mandated to assess cost-benefit as we suggested.

Our ability to enact such legislation with virtually no historical precedent, and the by and large positive way in which institutions have seriously attempted to abide by it bespeak the strength and resilience of democratic self-government. Such a legislation of consensus and mandated dialogue is, in my view, the best hope for continuing to develop the moral status of animals in society in a way that will change thought as well as behavior.

Notes and References

[1]*See* B. Rollin (1981), *Animal Rights and Human Morality* (Prometheus Books, Buffalo, NY), Part II. (Hereafter cited as *ARHM*).

[2]*Ibid.*

[3]*Animal Legal Defense Fund* v *The Department of Energy Conservation of the State of New York.* 1985. INDEX #6670/85.

[4]US Congress, Office of Technology Assessment (1986), *Alternatives to Animal Use in Research. Testing. and Education* (US Government Printing Office, Washington, DC), pp 310ff. (Hereafter cited as OTA).

[5]*See* J. Katz (1987), "The regulation of human experimentation in the United States—a personal odyssey," *IRB* **9**, 1–6.

[6]Animal Welfare Act, Section 13 (a), 6.

[7]OTA, p. 276.

[8]OTA, p. 286.

[9]P. Singer (1975), *Animal Liberation* (New York Review Press, New York, NY).

[10]*See ARHM*, Part I.

[11]*New York Times*, October 25, 1988, p. 1.

[12]*See ARHM*, Part III.

[13]B. Rollin (1989), *The Unheeded Cry: Animal Consciousness, Animal Pain and Science,* (Oxford University Press, Oxford, UK). (Hereafter cited as *UC*).

[14]*The State News*, (Michigan State University), February 27, 1989, p. 8.

[15]D. Rosenhan and M. Seligman (1984), *Abnormal Psychology* (Norton, New York, NY).

[16]OTA, p. 341.

[17]*See UC* and references therein.

[18]R. Kitchell and H. Erickson, eds. (1983), *Animal Pain: Perception and Alleviation* (American Physiological Society, Bethesda, MD).

[19]*See* J.W. Mason (1971), "A re-evaluation of the concept of 'nonspecificity' in stress theory," *Journal of Psychiatric Research* **8**, 323–333.

[20]*See* R. Kilgour (1978), "The application of animal behavior and the humane care of farm animals," *Journal of Animal Science* **46**, 1478.

[21]*See UC*, chapter 6.

[22]*See UC*, chapter 6.

[23]*See UC*, chapter 7.

[24]W. Russell and R. L. Birch (1954), *The Principles of Humane Experimental Technique* (Methuen, London).

[25]*See* Chapters 1 and 2 in B. E. Rollin and M. L. Kesel, eds. (1990), *The Experimental Animal in Biomedical Research* (CRC Press, Boca Raton, FL)

[26]*ARHM*, Part III.

[27]*Ibid.*

NIH Guidelines
and Animal Welfare

Lilly-Marlene Russow

Alphabet Soup:
NIH, PHS, APHIS, and so on

My objective in this paper is to begin to assess the adequacy of NIH regulations regarding the care and use of animals in research, education, and testing. In order to do so, we must first understand what those regulations are, and why and how they affect research involving animals.[1]

NIH guidelines are set out in the *Guide for the Care and Use of Laboratory Animals* (hereafter referred to as the *Guide*), but have almost nothing to say about actual experimental procedures involving animals. The *Guide* is almost exclusively concerned with animal care: cage size, sanitation, environment, and so on.[2] However, the *Guide* is the cornerstone of a broader policy on research involving animals. This policy, formulated by the Public Health Service (PHS) was endorsed by Congress in the Health Research Extension Act of 1985. Most of the regulation and administration of these regulations within PHS is handled through the Office for Protection from Research Risks (OPRR) at NIH.

Thus, the *Guide* itself is one element of a more complex regulatory system, and is best understood and evaluated in that

From: *Biomedical Ethics Reviews • 1990*
Eds.: J. Humber & R. Almeder ©1991 The Humana Press Inc., Clifton, NJ

context. According to this broader PHS policy, every institution receiving federal support or buying or selling animals across state lines must have an Institutional Care and Use Committee (IACUC), which, among other things, must develop a procedure to review all protocols—specific descriptions of the way in which an investigator proposes to use animals.[3] In this and other ways, PHS policy recognizes that concern for animal welfare must extend to the way the animals are used in the course of an experiment, and not be limited to the way they are cared for before or after the research—a position that would not be evident from looking at the *Guide* by itself. In the following discussion, I will, therefore, consider the entire PHS policy, including but not limited to NIH guidelines.

The other important component of official policy on the use of animals in research is the Animal Welfare Act, passed in 1966 and amended several times, most recently in 1989. It is administered by the United States Department of Agriculture (specifically the USDA's Animal Plant and Health Inspection Service—APHIS). I will also from time to time comment on the Animal Welfare Act (AWA); although technically this falls outside the scope of this paper, there are some important discrepancies between PHS policy and USDA regulations that are relevant to the evaluation of the former.

In the remainder of this paper, I shall begin with more theoretical concerns, and ask how current policies define the scope and limits of our obligations towards animals in research. From there, I shall move to the practical level, and discuss how these general theoretical premises translate into specific decisions about how, and even whether, particular experiments should be performed. In each case, I shall draw some conclusions about the adequacy of the policy. I shall, however, also end with a separate discussion of what else must be done, since many of the problems that remain unresolved could be addressed either at the theoretical or the practical level.

Theoretical Underpinnings:
Scope and Limits

Whenever a policy is adopted that requires us to engage in moral deliberation in a certain sort of way, that policy will involve important and possibly controversial presuppositions about the scope and limits of our moral obligations—what ought to be included in our assessment about the ethical permissibility of an action. Current guidelines about the use of animals in research are no exception. As with most regulations, however, the presuppositions are often left unstated, implicit in the policy but not spelled out. An adequate evaluation will require a careful analysis of those presuppositions; that will be the goal of this section.

As a philosopher, I am sorely tempted to frame the relevant presuppositions or underlying principles in terms of specific moral theories that might shape and justify our treatment of animals. Luckily, common sense prevails. The debate between utilitarianism and deontological theories is not likely to be resolved by legislation or at an IACUC meeting, and even within each of these two camps, there is no consensus on the implication of the theory with respect to our treatment of animals.[4] An effective policy must find a more accessible and practical framework within which to operate; if NIH guidelines are to be adequate, one must be able to discover the basic shape and scope of such a framework within these practical constraints implicit in the following three important realities about the way public policy decisions, as contrasted with philosophical conclusions are reached:

1. Whatever one thinks about the role of intuitions in developing a philosophically sophisticated moral theory, more specific deliberations in a real-world environment are very much a process of achieving reflective equilibrium between principles and (one hopes) carefully considered and objective intuitions about what one ought to do. Thus,

principles that directly clash with basic and very firmly held intuitions are likely to be rejected or greatly modified.

2. Because of (1), movement away from the status quo will be a slow process. As intuitions begin to be shaped by fairly modest principles, more reform becomes possible. Thus, issues that would have met with derisive rejection in 1966 or even 1984 (e.g., concern for the psychological well-being of primates) are now becoming part of the policy.

3. Intuitions do not come in two distinct flavors, utilitarian and deontological. Trying to shape our principles into one or the other pure form is, therefore, unlikely to be very helpful.[5]

The AWA and NIH guidelines are basically attempts to mediate between public support for scientific, especially biomedical, progress, and increasing sensitivity to the status of animals as living sentient creatures closely related and similar in morally relevant ways to us. Thus, we should not expect a radical moral theory with implications that diverge widely from the status quo; awakening sensitivity demands some change, but cannot start with a complete reversal of generally accepted positions. The basis for judgment will often be intuitions, whether carefully considered or immediate and emotional; as public policy rather than philosophy, these guidelines must be judged according to how well they make sense of such intuitions, and whether they can make those intuitions into an internally consistent policy that is also compatible with our broader ethical commitments. Above all, the underlying principles must be inferred from the policy; they are not explicit, not available simply to read off.

The most important component of current NIH policy, as contrasted with the situation in the 1960s and 70s, is also the most nebulous. Although it is never explicitly stated, the new policies adopted in 1985 definitely seem to presuppose that animals have an *independent moral status*. That is, some policy changes arguably make sense only if we assume that animals have a welfare independent of (or at least not limited to) their role as research subjects, and that to ignore the impact of animal use on that welfare is unjustifiable. Earlier regulations can all be adequately grounded by appeal to the need for good scientific practice—a diseased or stressed animal will not always yield reliable results—protection of those who work with the animals, or respect for the sensibilities of the general public.[6] Now, however, the presumption moves more towards a direct concern for the animal. For example, the investigator is expected to use anaesthetics and analgesics to relieve pain or distress, and the burden of proof is on him or her to justify their omission if the procedure is thought to require such abstention. Some of the AWA, especially Section 13, can be interpreted as moving in the same direction, although generally there is much more emphasis in the AWA on justification in terms of benefits to science and society.

All of this is buried too far under the surface to allow for much further elaboration or analysis. It is certainly clear that none of this is meant to accord animals a moral status equal to that of humans; PHS policy explicitly endorses the sacrifice of animals and their welfare for the improvement of human health. There is not even a move toward the possibly less rigid utilitarian demand that the amount of human suffering that is avoided or alleviated must outweigh the animal suffering caused by the research—what Singer calls the demand for equal moral consideration, as opposed to equal moral status.[7]

The difference between equal consideration and equal status is important: the claim that animals have equal moral standing seems to mean that there is no difference between the moral worth of a human life and that of an animal, and I know of no philosopher who has defended this position.[8] The claim that animals are entitled to equal moral consideration means only that if action A causes morally relevant harm to an animal, and action B causes the same amount (and possibly kind) of harm to a human, those harms must be given equal weight in our moral deliberation. It leaves open the very real possibility that action "A" might cause more harm when done to one sort of being rather than another. If killing a human being causes more harm than killing a rat, then equal consideration does not entail equal moral standing.

Whatever moral status is implicitly recognized does not seem tied to some characteristic of the beings to which it is granted, and the result appears arbitrary. The *Guide* describes itself as concerned with "any warm-blooded vertebrate animal used in research, testing, and education;" it notes without elaboration that the same humane principles are applicable to cold-blooded animals (cold-blooded vertebrates), but excludes farm animals that are not used for specifically biomedical research (that is, livestock used in research about production methods are not covered). It does not, however, explain why warm-bloodedness is associated with some morally relevant characteristic.

The AWA uses similar wording in describing its scope, but the USDA has explicitly excluded mice, rats, birds, and farm animals from coverage under the AWA,[9] again without giving any justification for this restriction. In order to resolve these inconsistencies and to avoid arbitrary and unjustifiable discrimination, NIH guidelines still need to provide an explicit, objective, and more reflective basis for recognizing moral status.

Nonetheless, the premise that animals have a significant independent moral status, but not equal standing (i.e., the same moral worth as humans) fits the constraints mentioned at the beginning of this section. It is widely accepted in society at large, is consistent with many peoples' considered and objective intuitions, and can be defended on plausible philosophical grounds. Further, the notion of independent but unequal moral status might be extended beyond the human/nonhuman animal split. Certainly in most people's minds, there is an implicit moral hierarchy under which a chimp has a higher moral status than a dog, which in turn is "worth more" than a mouse, and so on down the scale to cold-blooded animals, invertebrates, and single-celled animals, and this attitude is necessarily reflected in current guidelines. Although a full exploration of this last point would take us far beyond the scope of this paper, the basic premise is bounded on the one hand by a general conviction that any experience reasonably classified as pain or distress must be counted as intrinsically bad, something that we have a moral obligation to minimize, and on the other hand a belief that increasingly sophisticated and abstract cognitive abilities must somehow be factored into our moral scheme of things.[10]

There are, admittedly, several important theoretical issues still to be addressed: How does one handle so-called marginal cases—humans whose cognitive capacity is much lower than that of a normal dog or laboratory rat?[11] Which cognitive abilities are relevant, and why? Do we really have a one-dimensional hierarchy, or are several competing factors at work in our (justified?) preferential regard for some creatures over others? Is our special preference for dogs (as contrasted with the more intelligent pig) merely arbitrary, a social prejudice, or is there some theoretical justification for these discriminations?

It may turn out that careful philosophical analysis will show that such hierarchies are unjustified. However, given the first two constraints noted at the beginning of this section, practical realities of the current situation suggest that only a framework that posits differential moral status will achieve reflective equilibrium with generally held intuitions and attitudes, and hence any changes must be gradual. Nonetheless, more clarification is needed on this topic, and public policy that treats different species differently must at least be tied explicitly to criteria that might be morally relevant.[12]

A second key element of the framework within which federal policy is cast is the increased sensitivity to the ways in which animals may be harmed, and the moral significance of that harm. Many scientists have reservations about the term "animal suffering,"

perhaps thinking that it might imply a higher level of self-con-
sciousness or awareness than they are willing to attribute to most
nonhuman animals. These reservations are mirrored in the wording of federal policies: to my knowledge, the phrase "animal
suffering" is never used. Nevertheless, the sorts of harm addressed
in the *Guide* and other regulations go beyond acute physical pain,
and include other forms of stress and distress. Particularly for
primates, the newest version of the AWA requires special attention to psychological well-being, and in 1989, the National Research Council was directed to begin work on a set of guidelines
for recognizing and alleviating pain and distress in laboratory
animals.[13] Finally, PHS policy notes that "Unless the contrary is
established, investigators should consider that procedures that
cause pain or distress in human beings may cause pain or distress
in other animals."[14] Although "should consider" and "*may* cause
pain and distress" water down the force of this statement, it is a
valuable first step in establishing a better understanding of the
many ways in which an animal might suffer.

It is appropriate at this point to introduce the concept of the
"ethical cost" of a research project; the amount of morally relevant harm that is produced compared with the moral benefit
achieved.[15] Thus, if two experiments achieve the same results—
e.g., comparing the efficacy of a new anaesthetic with a well
understood one—but one involves the infliction of severe pain
whereas the other relies on tail-flick responses, the latter has a
much lower ethical cost. By the same token, two experiments
involving the same procedure may produce different benefits (a
new treatment for diabetes, or a new flavor for soda), and here
too, the ethical cost is greater if the same amount of morally
relevant harm, including animal suffering, produces a lesser benefit. It is a tautology that one ought to lower the ethical cost of any
action as much as possible, which is why the "Three Rs" of animal
research[16] are so widely endorsed (in word, if not always in action).

It would seem equally tautological that if the notion of ethical cost is to have any real meaning, there must be some experiments whose costs are judged to be too high, whose potential benefits do not outweigh the amount of harm caused. The issue is not whether there is some procedure that we would simply never allow, no matter what benefits might accrue, but the less controversial proposal that there are experiments that ought not to be done because the potential benefits are too small or trivial in comparison to the harm caused.

This last point is not motivated by specific examples that might strike even (or especially) a casual observer as clearly horrifying—causing paralysis in both hind legs of a cat or any of the other examples that are so easy to find in various advertisements and solicitation for donations. Rather, it follows necessarily from the theoretical commitments we have just been considering. If the ethical cost assigned to an animal's suffering is so minimal that any reason, including idle curiosity, the need to get another publication, or the desire to attract more attention at an elementary school science competition, can outweigh it, the notion of ethical cost which was endorsed in principle becomes totally meaningless.

However, there are still some ardent defenders of animal research who are unwilling to endorse, even in principle, the idea that there are some experiments that simply should not be done because their ethical cost is too high. Many more scientists and even some IACUC members agree in principle that some experiments cannot be justified, but find themselves unable to cite or imagine a single realistic[17] instance of such an experiment. Unfortunately, the *Guide* and other regulations remain mute on this point. Several schemata have been offered as a way of reflecting all aspects of the ethical costs of an experiment,[18] but they have not been incorporated into NIH policy. Failure to address the issue of ethical cost is arguably one of the main shortcomings in current policy.

Although evaluating the ethical cost of an experiment might seem unobjectionable, it requires a change in policy that many scientists resist: the consideration of scientific merit in the course of an IACUC's deliberation. Assessing ethical costs requires a judgment about the potential benefits of a research program, and that judgment can only be made by asking two questions concerning scientific merit: Are the experiments well designed from the perspective of scientific methodology, and how important are the issues that are being addressed and how do they fit in with current research in the field? PHS policy is deliberately vague about whether IACUCs should consider the scientific merit of the proposals they review, but many IACUCs and concerned scientists have taken the stance that questions of scientific merit should be "out of bounds." This stance must be changed if the notion of ethical costs is to be preserved, and it is difficult to imagine how one could justify an experiment without mentioning ethical costs.

It is tempting to think of this talk of "ethical costs" as an implicit endorsement of utilitarianism, but it need not be restricted in this way. Even though a deontological or rights-based theory will not reach a moral evaluation solely on the basis of beneficial vs. harmful consequences, there will be situations in which it must balance one right or duty against another. Questions such as how to balance one individual's right to liberty against another's right to protection from deliberate threats may be more problematic for a deontological theory than they are for utilitarianism; which violation of rights is morally worse? Alternatively, if my duty not to interfere with the autonomous life of an innocent individual becomes inconsistent with my duty to save human lives whenever possible, the deontologist must be prepared to calculate the ethical cost incurred by my failure to carry out one or the other duty.

Even within utilitarianism, the ethical costs need not be limited to hedonistic considerations; a more sophisticated version may have other values to be maximized, and hence, other considerations that affect the ethical cost. Finally, other rights-based theories such as the one endorsed by Tom Regan are even more complicated in that they are "mixed" or "moderate;" they attempt to make room for consideration of consequences within a generally deontological framework.

The issue of ethical costs is closely related to that of conflicts between values, but the two are not identical. Ethical costs may involve tradeoffs only between positive and negative values of the same sort. If we adopt a classical utilitarian theory[19] for example, our calculation of ethical cost need not address the problem of conflict; our job will merely be to identify the option that maximizes happiness. However, if one adopts a version of utilitarianism that asks us to maximize more than one value (pleasure, fulfillment, and autonomy, for example) one must be prepared to deal with cases in which there is a clash between two or more different sorts of values, perhaps incommensurable. If we think that scientific knowledge and freedom from suffering are both intrinsically good in basic ways, we must also recognize that, in some situations, (any area in which the only effective research methods currently available require the use of animals, for example) one of these values can be maximized only at the expense of the other. Similarly, a deontological theory will (*pace* Kant) inevitably encounter situations in which one must choose the lesser of two evils, in which, for example, one must abrogate one right to protect a more basic one. An adequate ethical framework must make room for such situations, but current guidelines have nothing to say on this subject.

The inevitability of conflict is guaranteed by the frequent disclaimers, both in the NIH *Guide* and in the AWA, to the effect that:

> Nothing in the *Guide* is intended to limit an investigator's freedom—indeed, obligation—to plan and conduct animal experiments in accord with scientific and humane principles.[20]

This comment could be taken to suggest that any experiment, if it is made as "humane" as possible given the objectives of the investigator, should still be permitted, in effect making freedom to investigate a "trump" value which will automatically take precedence. Many IACUC members interpret the passage differently, and see it as requiring a judicious balance between scientific principles and principles of humane concern for animal welfare. Under this interpretation, there may well be cases in which obligation to prevent or minimize animal suffering may win out over a desire to conduct a particular experiment if it were impossible to conduct the experiment "in accord with . . . humane principles." Given the lack of any further elaboration about how the comparison is to be made, and the absence of some sort of rating system to help ascertain ethical costs more fully, IACUC members are left to make the decision without further guidance. Other elements of PHS policy are similarly laudatory but vague. Translation of such admonitions is dealt with by the other crucial component of the policy: the decision that, for the most part, the protection of animal welfare rests with an Institutional Animal Care and Use Committee (IACUC) whose responsibility it is to ensure compliance with the *Guide*, the AWA, and all other relevant Federal policies. Thus, PHS policy both implicitly and explicitly sets the following guidelines for IACUCs: they are to give moral weight to the welfare of the animal itself, as having an independent moral status; and they are to pay primary attention to pain and other sorts of distress as crucial components of the ethical cost of an experiment. However, the framework is deliberately open-ended; rather than, for example, try to anticipate and rule on specific issues (e.g., "No multiple survival surgeries"), the onus is placed on the IACUC to decide whether a particular protocol fits within the general framework. Thus, the *Guide* states:

> Multiple major survival surgical procedures on a single animal are discouraged. However, under special circumstances they might be permitted with the approval of the

committee. One situation in which multiple survival surgical procedures might be justified is when they are related components of a research project. Cost savings alone is not an adequate reason for performing multiple survival surgical procedures.[21]

This passage contains a well formulated general principle (although further explanation or justification would still help guide decisions in difficult cases) that the IACUC can apply, deciding in each specific case whether the multiple surgeries are a necessary part of the research. I shall presently argue that, for the most part, this is the right way of going about the task of evaluating research. Whether one has a utilitarian, rights-based, or mixed value theory, public policy should not attempt to identify absolute general prohibitions or licenses that can be justifiably codified; there will always be the possibility that the ethical cost of a procedure will be outweighed by the ethical gain, whether that be measured in terms of utilitarian benefits or rights and duties. Instead, protocols can be evaluated only on a case-by-case basis.

The claim that a detailed case-by-case analysis is preferable to broader prohibitions is reinforced by the observation that many meetings designed to help IACUC members refine their decision process (including several meetings sponsored by the Scientists Center for Animal Welfare) have found that working through very detailed case studies is an important tool. This is a far cry, however, from the very sketchy information often cited in the mass media on both sides of the debate. Citing procedures that sound agonizing without providing objective evidence of distress, detailed information about anaesthetics and analgesics, or the purpose of the research do little to encourage a careful assessment of ethical costs. On the other hand, general lists of "scientific advancements that were possible only by using animals" without considering *how* the animals were used, and how the research could have been conducted at a lesser ethical cost, are equally unhelpful. The moral here cannot be emphasized too strongly: one needs to understand all aspects of the experiment before one can assess its ethical cost, and hence, reach a rational decision about whether it is justified.

The adequacy of this general approach, therefore, will depend on two factors: the general principles—how clearly they are articulated, and how well they are justified—and the mechanism by which the principles are interpreted and applied, case by case, on the basis of the sort of detailed information mentioned above. As we have seen in this section, the general principles that underlie current policies are a good first step, well in keeping with public sentiment and general intuitions; however, they stand in need of more explicit formulation and justification. The mechanism by which these principles are used to generate decisions about specific cases is, first and foremost, the IACUC.

In effect, NIH guidelines tell the IACUC what to look at, what factors to consider, but is silent about what to do in cases where those factors point in different directions, or how to translate general principles into specific decisions. Thus, currently the ultimate success or failure of NIH guidelines to provide adequate consideration of animal welfare will rest with the way in which IACUCs are formed, how they reach their decision, and how they assure compliance. This shall be the topic of the next section.

The IACUC

As just noted, the IACUC is the vital intermediary between PHS policy and final decisions about how and whether an experiment should be performed, when an environment is acceptable, whether an animal is suffering too much. An IACUC must inspect all animal facilities that fall under NIH and USDA guidelines, must review all proposals for the use of animals in teaching, testing, or research at a given institution, provide education for those who work with animals, and issue periodical assurances to the OPRR (currently, once a year). As we have just seen, this requires taking very broad and sometimes incomplete guidelines, and translating them into decisions about specific cases. Given the framework we have been trying to uncover, this will involve considerations about the amount of pain and distress caused to the animal involved, the benefits that might accrue, and how these two factors balance out against one another. What sorts of credentials should members of such a group have? And what sorts of questions should they be considering? In what follows, I shall suggest that these two questions really should be collapsed into one, and that although NIH guidelines emphasize the first, it is more important to pay attention to the second.

We have seen that there are three major areas of concern that an IACUC must investigate when deciding whether to approve a protocol:

1. the effect on the welfare of the animals involved, and possible alternatives with a lower ethical cost;
2. the potential for benefit that might justify the imposition of those costs; and
3. how to resolve conflicts.

It was mentioned earlier that PHS goes beyond both the NIH *Guide* and the AWA in requiring an IACUC of at least five members. The committee must include a veterinarian, someone experienced in the use of animals in research, a nonscientist, and someone not affiliated with the institution.[22] To some extent, these required members are supposed to be particularly well suited to address these concerns.

The veterinarian, of course, is supposed to be in the best position to determine how a procedure will affect the animals being used, whether the animals are healthy and being cared for properly, whether anaesthetics and analgesics are being used appropriately, and whether euthanasia procedures are consistent with AVMA policy, as required by the *Guide*. In many ways, the veterinarian can and should be in the best position to act as "the advocate for the animals," since he or she typically has the most knowledge of how the animal's welfare is affected. In the past, some veterinary schools had discouraged attention to "anthropomorphic" concerns about psychological well being, and specialization in laboratory animal medicine was rare, but both these problems are gradually lessening. The veterinarian can also play a crucial role in education, both in designing training programs and in informal consulting with investigators about the most humane treatment in a specific case. Thus, the veterinarian can and should be the first to address the first of our three questions listed above: how will the animals be affected, and what can be done to minimize the harm.

The role of the animal researcher on an IACUC is less obvious. One might assume that a researcher would be uniquely qualified to assess the scientific merit of a proposal, but in reality, as we have seen, often that is not part of his or her function at all. If, as I urged in the previous section, public policy begins to demand that scientific merit should be a factor in moral evaluation, this may change. If it does, that entails new concerns about the makeup of the IACUC; research involving animals spans so many different disciplines that a researcher in one area may be little more than a novice when it comes to experiments in a very different field. But what about the scientist now, when judgments of scientific merit are not central to IACUC deliberations?

One factor that all people in this category can be expected to share is their conviction about the importance of scientific research. First, it is hard to imagine that any researcher who uses animals would disagree with the claim that some, possibly most, well planned research is important enough to justify the use of animals. Second, someone active in research is likely to be more aware than the outsider that immediate and dramatic applications are not the only or most common benefit of even very valuable research. Scientific advances come unpredictably, inevitably with many false starts and dead ends, and often in very small steps. Thus, to expect to judge the value of a bit of research by asking, "How many lives will this experiment save?" is based on a naive view of scientific method, a view that an active researcher is less likely to fall into. Researchers are also keenly sensitive to the value of what is sometimes termed "pure knowledge," knowledge pursued primarily for the sake of simply understanding more. Although nonscientists might freely admit that music and poetry are "good for their own sake," not because they save lives, at least some people are less likely to reach the same appraisal of scientific understanding. By including a researcher on the IACUC, PHS policy ensures that this value will not be overlooked, thus implicitly accepting it as a good to be protected. Finally, on a more pragmatic level, the inclusion of an animal researcher helps to break down the "us vs. them" attitudes that often infect attempts to raise the issue of moral obligations in science.

The next member to be considered is the so-called nonscientist. Informally, many committees have decided that either the nonscientist or the nonaffiliated member should be an "ethicist." The problem here, of course, is that it's not at all clear what an "ethicist" is. There is also disagreement about how closely an"outsider" (either the nonscientist or the nonaffiliated member) should be allied to the animal welfare/animal rights movement; some people feel that "a radical" will disrupt the process irreparably, whereas others believe that the outsider's role is to be an advocate for the animals. Finally, it is unjustified to assume that the two positions, that of the nonscientist and that of the nonaffiliated representative, should be that closely allied. Although it is usually the case that the nonaffiliated member is, in fact, not a scientist, the role of a member *qua* nonscientist may differ from the role he or she or some other member should be expected to fill *qua* nonaffiliated member. For all these reasons, the role of these two members, and hence, the qualifications they might bring to an IACUC, are very much in need of further clarification. Since written policies do not shed any light on this issue, we must fall back on speculation: what might such people contribute, and, in that light, what qualifications are particularly important?

In considering the role of the nonscientist, it is helpful to recall our conclusions about the rationale for including a researcher on the committee. One important task for the researcher, I suggested, was to ensure that the committee operated with an adequate understanding of scientific method, and an appropriate sensitivity to the value of the scientific enterprise, pure as well as applied. Although society at large respects and supports science, it has become clear that this does not amount to an unqualified, blind endorsement. Few people would claim (especially after World War II) that the pursuit of scientific knowledge will justify and license every conceivable experiment. If society at large puts limits on what it thinks science ought to be allowed to do, and if this attitude is rational, one would want that viewpoint represented on an IACUC. Presumably, the nonscientist is in a better position than a scientist to be objective about the scope of such limits.

If this speculation about the purpose of requiring a nonscientific member is correct, this person ought to be especially attentive to questions about the potential conflicts between science and other societal values. One cannot guarantee that will happen either by requiring a nonscientist as a member, or by specifying some more specific qualification. Here, even more than with the veterinarian and the researcher, if the intent of NIH guidelines is to ensure that several important factors, which should be considered in an IACUC's deliberation, are not overlooked, explicitly identifying those factors would be more effective than leaving it up to the individual to guess what his or her role should be. Given the current situation, it is no wonder that many of the nonscientific members find themselves asking "why am I supposed to be here?"

Once we have identified the responsibility of the nonscientist in this way, we can also reach a more informed conclusion about his or er position vis a vis "the animal rights" movement. American attitudes towards animal research are ambivalent; increased publicity about the Draize test and LD-50 tests have made the public more critical, but there is still widespread conviction that the use of animals is an acceptable price to pay for some sorts of benefits (witness the frequent calls for more research on cancer, AIDS, and so on). A nonscientist who strives to make judgments that may achieve reflective equilibrium with such intuitions cannot afford to embrace a fully blown antivivisectionist position.

In many ways, the "outside member" fulfills some of the same functions as the nonscientist. Two differences, however, are worth noting. Failure to comply with NIH guidelines can result in the institution's losing all federal research funding—a serious threat to a large research facility or university. It would not be surprising, therefore, to find an institution to "encourage" IACUC members to overlook violations; an independent member of the committee might be in a better position to act as a whistle-blower. Second, the outside member, like the researcher, can be perceived as a liaison to an interested constituency; in this case, not other researchers, but the general public. Thus, the inclusion of an outside member affords the opportunity to foster an atmosphere of openness, fairness, and cooperation.

I have tried to be explicit about the fact that I have been guessing at NIH's intentions in ascribing these functions to various committee members; written policies only tell us what categories the members should be drawn from. However, it does seem reasonable to expect that since an IACUC does have the enormous responsibility of interpreting the guidelines, it should make sure that all aspects of the general issue of the use of animals in research should be considered. That expectation, together with some reasonable assumptions about what various sorts of people might bring to the committee by way of special expertise or commitment, has shaped my inferences.

Ideally, though, such second-guessing should not be necessary. The guidelines themselves should say more about the scope of an IACUC's deliberation, rather than hoping that the required diversity itself will ensure that all relevant factors are given due weight. Until that is done, a duly constituted IACUC might well turn out to be the best forum for these deliberations, but it will always be possible to subvert the goal of effective decision-making by choosing people who fit the categories but do not bring to their work the expected expertise or commitment. It is possible, under current guidelines, to have an IACUC comprising a veterinarian who has never worked with rodents, a researcher who is interested only in snakes, and an outsider who is a USDA licensed dog dealer. As I noted at the end of the previous section, the guidelines emphasize a careful case-by-case approach, rather than attempting to legislate against certain practices *tout court*. No matter how good this approach is in principle, it makes it especially important to make sure that all relevant factors are weighed in the process.

I began this section by suggesting that questions about the credentials of IACUC members were less important than questions about what such a group should consider. The subsequent analysis highlighted three areas of particular concern:

1. a realistic and careful assessment of the animals' needs and welfare, together with the ability to predict the impact of various procedures on the animals;
2. an understanding of the scientific method as it applies to research involving animals, a commitment to the value of such research, and the ability to serve as a respected liaison with the scientific community; and
3. an ability and willingness to think about the appropriate role of scientific research in the larger context of societal values.

These are delicate and complex topics to which one cannot do justice with heavy-handed legislation, but one can do more to make sure that the IACUC will be an appropriate forum in which to deliberate about them and translate those deliberations into specific decisions.

Unfinished Business

Anyone who has served on an IACUC can easily generate a long list of unresolved disputes. The use of pound/animals, and the question of whether animals should be reused (reducing the total number of animals used, but putting more stress on individual animals) would certainly be near the top, but the list could be expanded indefinitely. Clarifying some of the vague, incomplete, or ambiguous aspects of current policy that I have attempted to identify in the previous sections of the paper might help resolve some of the disputes: for example, if we had a more fully worked out basis for claiming that there is a hierarchy of moral standing in which harming one sort of animal may be easier to justify than harming another, we would be better equipped rationally to decide whether the differences between pound dogs and purpose-bred dogs is morally relevant.[23] In other cases, however, conflict is inevitable. Anyone attempting to reach decisions about animal research will inevitably be pulled in two different directions in two different ways. The first source of tension has to do with the relation between legislation and morality: how far can or should we attempt to pass laws requiring that people behave morally, and when do we stop and say that some decisions must be guided by individual conscience? Second, how does one decide when the price of knowledge becomes too great—how does one reconcile science and humanity? There are some scientists who would claim that all decisions about research should be left to the individual investigators, and that the pursuit of scientific knowledge is among one of humanity's highest goods. People in the animal rights movement might counter by arguing that previous abuses prove that individual conscience is sometimes not sufficient, that animals who cannot protect themselves ought, like children in the same situation, to receive special protection under the law, and that scientific research is a two-edged sword, not an unmitigated blessing. As usual, we are likely to find the truth somewhere in between. It would be unfair as well as unrealistic to expect NIH to provide the final answer, but I hope to have suggested some ways in which it might more effectively occupy and map that territory in the middle that it has already started to explore.

Notes and References

[1]Much of the material in the following paragraphs, and other factual information about various regulations can be found in M.T. Phillips and J.A. Sechzer. (1989) *Animal Research and Ethical Conflict* (Springer Verlag, New York, NY).

[2]The main exception is to be found in Chapter 3, "Veterinary Care." The sections on anaesthesia and analgesia, surgery and postsurgical care, and euthanasia (pp. 37–39), will be relevant to some (but not all) experimental procedures as well as to the care of animals before and after their use as research subjects. I am also, for the moment, excluding consideration of Appendix D, since I will discuss the more general PHS policy shortly.

[3]Although PHS does not mandate the use of any one form for protocols, a good protocol review form will normally ask the investigator to comment on (1) why a nonanimal model would not work, (2) the choice of species and the number of animals used, (3) specific sorts of pain or distress that might be expected, (4) what anaesthetics or analgesics will be used to alleviate pain, and (5) qualifications and training of the investigators, technicians, and anyone else working with the animals or responsible for their care. Sample protocol review forms are available from OPRR.

[3]Peter Singer and Raymond Frey are both act-utilitarians, for example, yet the former condemns almost all research involving animals whereas the latter would claim that much of it is morally justified. *See* Peter Singer. (1975) *Animal Liberation* (Avon Books, New York, NY), and Raymond Frey. (1983) *Rights, Killing, and Suffering* (Basil Blackwell, Oxford, UK). Within the deontological tradition, we can

find the same disagreement between Kant and Regan. *See* Tom Regan. (1983) *The Case for Animal Rights* (University of California Press, Berkley, CA) for a discussion of his own view and Kant's.

[5]Cf. Lilly-Marlene Russow. (1990) "Ethical Theory and the Moral Status of Animals," *Hastings Center Report* **20** (**Suppl.**), 4–8.

[6]Many people have suggested a causal connection between the swift passage of the Animal Welfare Act and the publication of a very powerful article on pets being stolen and sold to research labs that was published in *Life* magazine earlier that year.

[7]Singer, *Animal Liberation*, p. 7; *see also* pp. 15–22.

[8]Singer explicitly denies "equal worth" on pp. 21–22 of *Animal Liberation*. Regan draws a similar conclusion in *The Case for Animal Rights*, pp. 324–325. Claims about equal worth are often attributed to "the animal rights movement", but sorting out that debate would take us too far afield of the topic at hand.

[9]*See* Phillips and Sechzer, p. 22.

[10]For a critical assessment of this belief, *see* S.F. Sapontzis. (1987) *Morals, Reason, and Animals* (Temple University Press, Philadelphia, PA) esp. pp. 129–137, and 159–161.

[11]There is a vast amount of literature on this topic, going back at least as far as Peter Singer, *Animal Liberation*, pp. 19–23. For a recent discussion of the topic, *see* James Lindemann Nelson, "Animals, Handicapped Children, and the Tragedy of Marginal Cases", *Journal of Medical Ethics*, v. 14 (1988), pp. 191–193.

[12]For a more detailed discussion of factors that might be morally relevant, *see* Morton, D. M., Burghardt, G. M., and Smith, J. A. "Critical Anthropomorphism, Animal Suffering, and the Ecological Context." *The Hastings Center Report*, pp. 17–18.

[13]Phillips and Sechzer, p. 238.

[14]NIH *Guide*, p. 82

[15]To be precise, a determination of ethical cost should also factor in the probability of the harm and benefit; a benefit that is only remotely possible will not weigh as heavily against an almost certain harm as much as a more likely gain would. However, the simplified version will be adequate for our purposes here.

[16]The "Three Rs" are "Reduce" (reduce the number of animals used), "Refine" (modify the experiment so as to cause less pain, distress,

or other harm), and "Replace" (replace live animals with tissue cultures, computer models, videotapes [for education], and other alternatives).

[17]The qualification "realistic" is important here. People who fit in this category are usually willing to agree that we shouldn't allow someone to use a blowtorch to cause third degree burns over 60% of a guinea pig's back "just to see how many survive," but when the proposal is suitably dressed up with typical academic claims about the need to study severe shock reactions, the reluctance to "just say no" reemerges.

[18]*See*, for example, Patrick Bateson. (1986) "When to Experiment on Animals," *New Scientist*, **20**, pp. 30–32; Canadian and some European regulations also require this sort of evaluation.

[19] I.e., a version of utilitarianism that is (a) more or less in the "act-utilitarian" camp, and (b) is basically hedonistic. Jeremy Bentham is, of course, the best example.

[20]Some IACUC members interpret this to mean that an IACUC should never refuse to allow a procedure if the investigator insists that it is the most humane method of achieving whatever scientific result is being pursued.

[21]NIH *Guide*, pp. 9–10.

[22]The same person can fulfill more than one role; the nonscientist might also be the "outsider." Most IACUCs are likely to have researchers make up the vast majority of the committee.

[23]The dispute alluded to here has arisen in numerous communities. Many people object to the practice of making animals from municipal pounds or shelters available for use in the laboratory, whereas the opposing side argues that if the animals are to be euthanized anyway, it would be better to use them in research, rather than letting them die uselessly while breeding other animals specifically for research.

Author Index

Subject Index

A

AZT, 125, 153, 154
African Americans, 31, 38, 45
American Academy
    of Pediatrics, 63
American Medical Association,
    14, 63, 136
American Medical News, 150
American Physiological
    Society, 209
American Public Health
    Association, 106, 118
American Veterinary Medical
    Association, 212, 243
animal consciousness, 219–221
Animal Plant and Health
    Inspection Service, 230
Animal Rights Movement, 202
Animal Welfare Act, 191,
    196–200, 209–221, 215,
    220–221, 230–250
Australia, 53

B

Blue Cross/Blue Shield, 132

C

Canada, 29, 35, 37, 53, 54, 94,
    99, 104, 121, 133, 148,
    157, 174
Center for Animal Welfare, 241
Chrysler, 62
Consumer Choice Plan for the
    1990s, 107

D

Department of Health and
    Human Services, 210
Dole Brown Bill, 192, 209

E

ESRD, 154
egalitarianism, 44, 45, 47, 55
Employer provided health
    insurance, 29

F

FDA, 147, 212

255